# Chapter 1

# FRUIT IN THE GARDEN

Botanists know exactly what they mean when they talk about 'fruit'. A fruit for them is the seed-bearing organ of a plant. Luscious pears and juicy oranges are of course included – but so are grains of wheat, walnuts and dried peas!

For the gardener the word 'fruit' has a much more restricted and vaguer meaning – and there is no general agreement where the classification should start and finish. All fleshy seed-bearing organs which are either cooked or eaten fresh for dessert are included. But do you include nuts (which are not fleshy) and rhubarb (which is not truly a fruit)? It's up to you, but they are not included in this book.

Some fruits are traditionally picked from hedgerows – elderberries, sloes, blackberries and hips of shrub roses. For practical purposes such sources can be ignored – if you want fruit without paying for it, you will have to grow your own.

Growing fruit in the garden is decorative, interesting . . . and rewarding when you consider shop prices. It is therefore not surprising that the story of fruit growing in the garden is as old as gardening itself. What is surprising is the sophistication of the subject in those early days. In the 16th century numerous varieties of apples grafted on to crab apple rootstocks were grown in British gardens and there were 'Peares of all sortes'. In addition the textbooks of the day described white and red peaches, cherries, bullaces, quinces, raspberries, white and red strawberries, currants, mulberries, medlars, bilberries, grapes, gooseberries, damsons, apricots, plums . . . a list which would do credit to a specialist catalogue of today.

These Elizabethan books also bemoaned the fact that British gardeners did not grow as many fruit trees and bushes as continental gardeners – a criticism which could probably be made to this day. Only one UK garden in three grows fruit, and there are two basic reasons for this lack of universal appeal. First of all, there is the view that fruit growing is difficult and time-consuming. This, of course, is not true – once planted and established the fruit garden gives more for less effort than the vegetable plot. Perhaps some textbooks are to blame – descriptions of pruning often make it sound as complex and delicate as brain surgery.

Secondly, there is the view that fruit growing is space-demanding. Visions of spreading trees, ladders for picking and the thought of room-sized fruit cages deter the gardener with a pocket-handkerchief plot, but fruit trees these days need take up no more space than rose bushes.

Things are now changing, and in recent years there has been an upsurge of interest in growing fruit, both in the ground and in containers. There are a number of reasons for this increase in domestic fruit culture. The garden centre with its container-grown plants has enabled trees, canes and bushes to be seen in flower and in fruit, and these containers are available for planting all the year round. In addition the introduction of dwarfing stocks has meant that apple, pear, plum and cherry can be grown as trees in small gardens. Catalogues are so much more exciting these days – there are new types of fruit such as the kiwi fruit and Cape gooseberry, plus new growth forms such as the 'patio' fruit trees.

Fruit should be grown in every garden. A separate fruit garden is ideal, of course, but you need both space and money for such a feature. But soft fruit is quite at home in a container or mixed border, and non-vigorous tree fruit can be grown in pots or as specimen plants in the lawn or shrub border.

The great fascination is to be able to pick the produce in peak condition straight from the tree or bush. Another plus point is that you can grow delicious varieties which aren't found in the shops.

An introductory section should end on a high note with words of encouragement and enthusiasm, but this one ends with a note of caution. Planting a fruit tree or bush is a long-term investment, which means that you should read the appropriate chapter carefully *before* looking through the catalogue. Then choose carefully, making sure that both fruit type and variety are suitable for your particular conditions. For example, few varieties of pears are suitable for northern and Midland gardens despite the glowing praise for them in some catalogues.

Never guess that a variety will be suitable – do check. You might like *Cox's Orange Pippin, Golden Delicious* or *Granny Smith* based on shop-bought experience, but none of these apple varieties is a good choice for the garden. You might think that the compact tree in the garden centre won't be a problem in your small garden, but do check the rootstock before buying. You might think that the raspberry canes look healthy, but check that they are certified virus-free. As stated earlier, fruit growing in the garden can be decorative, interesting and rewarding, but it can also be disappointing if you blunder into it without knowing what you are doing. The purpose of this book is to guide you along the right road.

# Choosing the Right Type

You will see some plants labelled with the letters AGM after the name; this shows that the variety holds the RHS Award of Garden Merit. This indicates that the plant is recommended by the Royal Horticultural Society, having been selected after extensive trials and inspections by a panel of experts as being the best of its kind.

Fruit-bearing plants come in all shapes and sizes. In the right setting, a stately apple tree with stout and spreading branches is not out of place in a large garden – nor is a row of alpine strawberries nestling under the roses. To make some sense out of this vast array of trees, shrubs and canes a variety of classifications have been proposed. Set out below is one of the simpler ways of dividing up the various types of fruits which are grown in Britain.

## TREE FRUIT
(other name: top fruit)

Included here are the larger and stouter fruit-bearing plants. Most of them adopt a tree form (a plant with a single main stem) in their natural state.

Fruiting does not take place for several years after grafting or germination, but once established the fruiting season usually lasts longer than with soft fruit. In addition the yield per plant is usually much higher: 4.5–180 kg (10–400 lb) per tree compared with 0.25–11 kg (½–25 lb) per plant from soft fruit.

Most tree fruits belong to the rose family.

### Pome tree varieties
Hardy trees which bear fleshy fruit which have a central cavity containing several small seeds (pips):
**Apple**
**Pear**
**Medlar**

### Other outdoor varieties
Hardy fruit-bearing trees which are either common in the countryside or uncommon in the fruit garden:
**Mulberry**
**Elderberry**
**Quince**
**Sloe**

### Stone tree varieties
Hardy trees which bear fleshy fruits which have a central cavity containing a large hard seed (stone):
**Plum**
**Cherry**

### Tender varieties
Frost-sensitive plants which can be grown outdoors in sheltered localities in southern districts, but are much more successful when grown under glass:
**Apricot**
**Peach**
**Nectarine**
**Fig**

## SOFT FRUIT

Included here are the smaller fruit-bearing plants. They require less space than most types of tree fruit and they bear fruit quite soon after planting, but their productive life is shorter.

Pruning is usually easier than with tree fruit, but they do need more protection from birds.

### Bedding varieties
Hardy non-woody plants which bear fruit within a year if planted in late summer:
**Strawberry**

### Cane varieties
Hardy woody plants bearing long and slender shoots. They bear fruit in the second year after planting:
**Raspberry**
**Blackberry**

*Hybrid berries – crosses between raspberry, blackberry and/or dewberry:*

| | |
|---|---|
| **Boysenberry** | **Silvanberry** |
| **Dewberry** | **Sunberry** |
| **Hildaberry** | **Tayberry** |
| **Japanese wineberry** | **Tummelberry** |
| **King's Acre berry** | **Veitchberry** |
| **Loganberry** | **Youngberry** |
| **Marionberry** | |

### Bush varieties
Hardy woody plants bearing spreading fruit-bearing branches. These are not all cut away after cropping:
**Gooseberry**
**Worcesterberry**
**Blackcurrant**
**Red and white currant**

*Heathland berries – require very acid soil:*
**Lowbush blueberry**
**Highbush blueberry**
**Cranberry**

### Tender varieties
Frost-sensitive or warmth-loving plants which require greenhouse culture in the north and Midlands:
**Cape gooseberry**
**Grape**
**Kiwi fruit**
**Melon**

# The NEW FRUIT EXPERT

## Dr D. G. Hessayon

Published by Expert Books
a division of Transworld Publishers

Copyright © Expert Publications Ltd 2015

The right of Dr D. G. Hessayon to be identified as author of this work
has been asserted in accordance with sections
77 and 78 of the Copyright Designs and Patents Act 1988.

A catalogue record for this book is available from the British Library

TRANSWORLD PUBLISHERS
61–63 Uxbridge Road, London W5 5SA
a division of the Random House Group Ltd

# CONTENTS

**Acknowledgments**

The author wishes to acknowledge the painstaking work of Gill Jackson, Linda Fensom and Constance Barry. Grateful acknowledgement is also made for the help or photographs received from Angelina Gibbs, Joan Hessayon, Jacqueline Norris, Jane Llewelyn, the House of Heyes, Crown Copyright: Brogdale EHS, Garden World Images Ltd, Pat Brindley, Richard Cumberland, Michael Warren and A–Z Collection.

John Woodbridge provided both artistry and design work. Mike Standage prepared most of the paintings for this book. Other artists who contributed were Norman Barber, Bob Bampton, Richard Bell, David Thelwell and Christine Wilson.

For this revised edition the publishers would further wish to thank Val and Steve Bradley, Chris Bradley, Brenda and Robert Updegraff, Peter Seabrook, Nick Dunn, Will Sibley, Janet Shuter, Frank P. Matthews Trees For Life®, Garden World Images Ltd and Marshalls Seeds.

The Random House Group Ltd supports the Forest Stewardship Council (FSC®), the leading International forest-certification organisation. All our titles that are printed on Greenpeace-approved FSC®-certified paper carry the FSC® logo. Our paper procurement policy can be found at www.rbooks.co.uk/environment

Reproduction by Spot on Digital Imaging Ltd, Gomm Road,

# Pest and Disease Control

## DON'T TRY TO KILL EVERYTHING

Not all insects are pests – many are positive allies in the war against plant troubles. Obviously these should not be harmed and neither should the majority of the insect population – the ones which are neither friends nor foes. There will be times when plant pests and diseases will attack, but even here small infestations of minor pests can be ignored (e.g. cuckoo spit) or picked off by hand (e.g. caterpillars, rolled leaves and foliage damaged by leaf miner).

*⅙ in. brown beetle*

*Pests like the raspberry beetle will damage flowers and lay eggs that mean the next generation can spoil developing fruits. Knowing when to spray can be critical – see page 107.*

## SPRAY IF YOU HAVE TO

Spraying is called for when an important pest is in danger of getting out of hand. Pesticides are safe to use in the way described on the label, but you **must** follow the instructions and precautions carefully. A wide range of controls is offered by most garden shops – look at the labels carefully before making your choice.

The front will tell you whether it is an insecticide, a fungicide or a herbicide. Make sure that the product is recommended for the plants you wish to spray. If it is to be used on fruit or vegetables, check that the harvest interval (the waiting period between the last spray application and harvesting edible crops) is acceptable. Do not make the mixture stronger than recommended.

When spraying, choose a calm, dry day with good light, but not bright sunshine. Use a fine, forceful jet and spray thoroughly until the leaves are covered with liquid and it is just beginning to run off. Do not spray open delicate blooms.

After spraying, wash out the equipment thoroughly, and wash your hands and face. Do not keep any spray solution you have made up until next time, and always store packs in a safe place. Do not keep unlabelled packs. Dispose of empty containers safely.

## INSECTICIDES

Insecticides are products that kill insects and/or other pests. As noted below, there are various types of active ingredients and these can act in one, two or all three different ways. If you are new to gardening, read these notes before you buy any products. There are dust formulations, such as ant killers, but these powder formulations are not popular as they leave a white deposit. Aerosols are also available, but the most popular type of insecticide you are likely to find is a liquid spray. Traditionally the concentrate in the bottle is diluted before being poured into a hand- or pressure-sprayer, but a ready-to-use (RTU) spray gun is much more convenient if you have only a small area to treat.

### Insect-contact types

These work by hitting and killing the pests, and are used against sap-sucking insects such as aphids and capsid bugs. They have no activity against pests that arrive after spraying, so it is necessary to spray when insects are seen and not before they attack. Use a forceful jet and cover all parts. Examples – insecticidal soap, pyrethins. Some act as poisons while others have a physical mode of action, either by blocking the breathing pores or gumming the pests to the leaf surface. Examples include plant oils and urea/foliar lattice.

### Systemic types

These work by going inside the plant and then moving in the sap stream – they are used against sap-sucking insects and some caterpillars. They do have activity against pests that arrive after spraying; in addition, they kill aphids hidden from the spray. New growth is protected. Complete cover is not essential. Examples – thiacloprid, acetamiprid.

### The active ingredients

Many of the products on the garden-centre shelves are **chemical insecticides**, which are modern complex materials developed in the laboratory and manufactured by large chemical companies. They are safe to use as directed, but the number of active ingredients has been greatly reduced in recent years; as a result very few systemic insecticides remain. At the same time, there has been an increase in the number of **organic insecticides** – fatty acids, insecticidal soap, rape seed oil and pyrethins are available. These are derived from plants and are generally less effective than their chemical equivalents; they need to be applied more often to achieve the same levels of control, but they are acceptable to organic gardeners. For **biological insecticides** you will have to find a specialist supplier – these products are based on living organisms (nematodes, bacteria, etc.) which are natural enemies of the pest to be controlled. They may be reasonably successful under glass but less so outdoors.

5

# FUNGICIDES

These are products which are used to control diseases caused by fungi. They have no effect on the cankers, leaf spots and rots caused by bacteria, nor are they of any use against the many viral diseases that can afflict garden plants. Keen gardeners spray their roses and fruit, but fungicides are less widely used than insecticides. There are two basic types, and timing is all-important. When to put on the first spray depends on the type you choose and the nature of the problem, so it is necessary to read the label carefully. This should also indicate how often a crop can be sprayed and list any crops or varieties that should not be treated with this particular product.

## Preventative types

These work by covering the plant with a protective coat which kills the fungal spores that arrive after spraying. Ideally, the first spray should go on before the disease has started, but in practice the initial treatment usually takes place when the first spots are seen. Repeat as instructed. In some cases (e.g. peach leaf curl) the first spray has to be applied before the disease is seen. Examples – copper and sulphur.

## Systemic types

These work by going inside the plant and then moving in the sap stream. Protection is better than with a preventative type, as areas missed by the spray are reached. There may be a minor curative effect on small disease spots, but they will not clear up a bad infection. Repeat as instructed. In a few cases (e.g. mildew and scab) the first spray has to be applied before the disease is seen. Example – myclobutanil.

## The active ingredients

Most of the products on the shelves are **chemical fungicides**, which are modern complex materials developed in the laboratory and manufactured by large chemical companies. The choice these days is very limited, but the ones that are available have been thoroughly tested and are considered by government to be safe to use, but only on those food crops specified on the label. Even with permitted crops you may find that there is a harvesting interval: i.e. a waiting period between the last spray application and harvesting edible crops. You will also find several **'green' fungicides** offered for sale – these are based on simple, age-old remedies such as copper compounds. They are usually less effective than the chemical fungicides, but they are generally acceptable to organic gardeners despite the non-organic nature of some of them. The interval before harvesting is one day or none at all.

*If fungal diseases are not treated they can cause the tree to weaken gradually and eventually it can stop growing and fruiting.*

# Chapter 2

# TREE FRUIT

The tree you propose to plant will bear fruit for 20–60 years, depending on the type and location. You can choose one which will grow knee-high or as tall as a house – the factors again are type and location. Obviously the choice is vast, so get to know the basic facts before you buy.

The usual growth pattern is the **free-standing tree** – here you will find a central trunk with a crown of branches radiating in all directions. Where space is limited a **supported tree** can be grown – see page 12 for details.

The eventual height of the tree will be determined by type, growing conditions and by the rootstock on which the variety has been grafted. The growth pattern is determined by pruning and training – details are given in the instructions for individual fruits.

Pick the sunniest spot available – full sun and a mild climate are essential for the tender types. The soil should be reasonably deep, not prone to waterlogging and ideally it should be slightly acid (pH 6.5). If the site is exposed, you will have to provide some protection from strong winds – see page 33. Frost can be a serious menace at blossom time – if you live in a frost-prone area, choose a late-flowering variety or one with high frost tolerance. Planting a supported tree against a south-facing wall saves ground space as well as helping in the battle against frost. This method of cultivation should be used for the tender types of tree fruit.

Having chosen the fruit you wish to grow and the spot where it will live (remember the effect of its shadow on other plants when picking the site), it is now time to choose the variety. The usual advice is to go for dessert (eating) varieties rather than culinary (cooking) varieties where space is short. Most tree fruits require a pollinating partner to ensure satisfactory fruit set. You will have to plant a partner if fruit trees are not common in your area – seek advice or read the appropriate section in this book. Where several varieties are to be grown, pick ones which will provide an extended cropping period.

So choose your fruit tree from the following pages. The tender ones may only be for the lucky few, but the hardy types should succeed almost anywhere apart from upland and very wet localities.

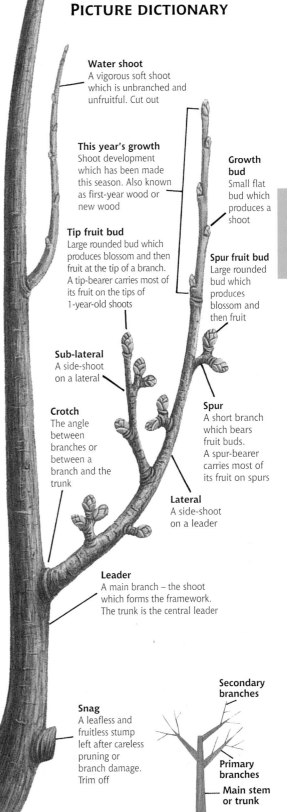

## PICTURE DICTIONARY

**Water shoot**
A vigorous soft shoot which is unbranched and unfruitful. Cut out

**This year's growth**
Shoot development which has been made this season. Also known as first-year wood or new wood

**Growth bud**
Small flat bud which produces a shoot

**Tip fruit bud**
Large rounded bud which produces blossom and then fruit at the tip of a branch. A tip-bearer carries most of its fruit on the tips of 1-year-old shoots

**Spur fruit bud**
Large rounded bud which produces blossom and then fruit

**Sub-lateral**
A side-shoot on a lateral

**Spur**
A short branch which bears fruit buds. A spur-bearer carries most of its fruit on spurs

**Crotch**
The angle between branches or between a branch and the trunk

**Lateral**
A side-shoot on a leader

**Leader**
A main branch – the shoot which forms the framework. The trunk is the central leader

**Snag**
A leafless and fruitless stump left after careless pruning or branch damage. Trim off

**Secondary branches**

**Primary branches**

**Main stem or trunk**

# Buying

You should buy fruit trees only from a supplier whom you know to be reputable – an often quoted (and often ignored) piece of advice. Do avoid mail-order bargains from unknown nurseries – if you do buy through the post then order from a company with a good reputation and put your order in early. The widest selection of varieties is available in autumn. By early spring some of the old favourites and new introductions are sold out.

If possible go along personally to buy your plants so that you can discuss your choice. Make sure that you find out the type of rootstock used – see the appropriate sections of this book for recommended stocks.

There are no hard and fast rules for the best age of trees for planting. If you really know about pruning then choose a 1-year-old feathered maiden – you will save money and you can be sure that it will be properly trained. On the other hand you should buy a 2- or 3-year-old tree if you have little or no practical experience – the basic framework will have been created by the nurseryman. These trained and partly trained trees are the type usually on offer – 1-year-old maidens are stocked by only a few garden centres.

The tree you buy will have been dug up at the nursery and damp-absorbent material packed around the bare roots before transportation – a bare-rooted tree designed for planting in the dormant season. You will have to buy a container-grown plant if you have missed the traditional November–March planting season. Look at the container carefully before you buy – see below. Give it the tug test. Lift the tree by holding the base of the main stem and pulling gently upwards. Don't buy the tree if the soil ball readily moves out of the container.

## PLANTING MATERIAL

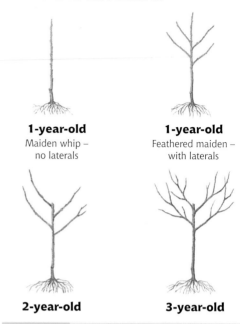

**1-year-old**
Maiden whip –
no laterals

**1-year-old**
Feathered maiden –
with laterals

**2-year-old**

**3-year-old**

| 1-year-old 'maiden' | Untrained. You will have to prune for about 3 years to produce a satisfactory framework of branches |
| --- | --- |
| 2-year-old | Partly trained. You will have to continue training to produce a satisfactory framework |
| 3–4-year-old | Trained. Purpose of pruning will be to maintain balance between growth and fruitfulness |
| Over 4-year-old | Generally too old for planting. Establishment may be very slow |

## THE THINGS TO LOOK FOR

### Container-grown tree

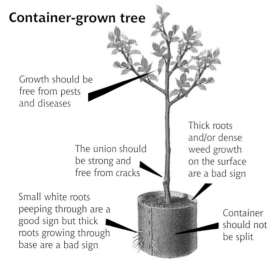

Growth should be free from pests and diseases

The union should be strong and free from cracks

Thick roots and/or dense weed growth on the surface are a bad sign

Small white roots peeping through are a good sign but thick roots growing through base are a bad sign

Container should not be split

### Bare-rooted tree

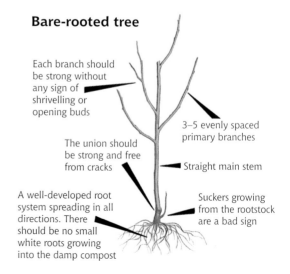

Each branch should be strong without any sign of shrivelling or opening buds

The union should be strong and free from cracks

A well-developed root system spreading in all directions. There should be no small white roots growing into the damp compost

3–5 evenly spaced primary branches

Straight main stem

Suckers growing from the rootstock are a bad sign

# Planting

Having bought a good plant by following the guidance on the previous page, you must now give it a good home. Traditionally, it was unacceptable to plant in a spot which had recently housed the same type of fruit, but recent research has shown that treating the roots of the new tree (or soil used to fill the planting hole) with a preparation of beneficial mycorrhizal fungi which colonize the new roots and aid the plant's growth mean it is possible to plant trees of the same type where they have been grown previously.

Planting of bare-rooted plants should take place while they are dormant, between late autumn and early spring. The preferred time is November whilst the soil is still warm. Soil conditions are as important as the calendar – the ground must be neither frozen nor waterlogged.

## GETTING THE SOIL READY

You should dig the site at least a month before the proposed time of planting. Double digging is recommended to aerate the topsoil and break up the subsoil. See *The Garden Expert* (page 12) for details. If the trees are to be widely spaced there is no need to dig the whole area. Dig a 1 m (3 ft) square plot where each planting site is to be – remove all perennial weeds during this soil preparation. For closely planted trees dig the whole length of the area to be planted.

Most soils can be successfully prepared in this way – do add plenty of organic matter if the soil is sandy or heavy. Very heavy clays and shallow chalks can be a difficult problem – the only satisfactory plan here is to add topsoil rather than trying to double dig.

## PUTTING IN THE STAKE
### Bare-rooted tree

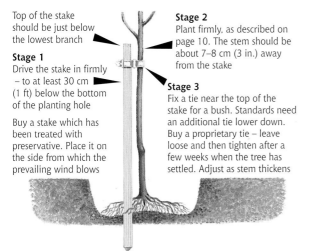

Top of the stake should be just below the lowest branch

**Stage 1**
Drive the stake in firmly – to at least 30 cm (1 ft) below the bottom of the planting hole

Buy a stake which has been treated with preservative. Place it on the side from which the prevailing wind blows

**Stage 2**
Plant firmly, as described on page 10. The stem should be about 7–8 cm (3 in.) away from the stake

**Stage 3**
Fix a tie near the top of the stake for a bush. Standards need an additional tie lower down. Buy a proprietary tie – leave loose and then tighten after a few weeks when the tree has settled. Adjust as stem thickens

### Container-grown tree

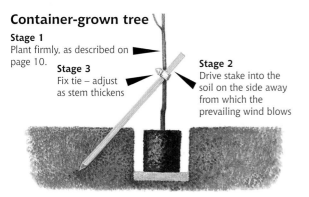

**Stage 1**
Plant firmly, as described on page 10.

**Stage 3**
Fix tie – adjust as stem thickens

**Stage 2**
Drive stake into the soil on the side away from which the prevailing wind blows

## GETTING THE TREE READY

If planting has to be delayed for a few days keep the tree in an unheated but frost-free garage or shed. Do not remove packaging around roots until planting day

Plunge roots in a bucket of water for about 2 hours if they appear dry or if the stems are shrivelled. Roots must never be allowed to dry out before planting – keep them covered until you are ready to set the plant in its planting hole

Cut back any damaged or very long roots to about 30 cm (12 in.)

## DIGGING THE HOLE

The first step is to mark out the planting stations with canes to make sure that the trees will be spaced out as planned. Next, the planting hole for each tree must be dug, and the commonest mistake is to dig a hole which is too deep and too narrow to house the roots properly. Another mistake is to try to turn tree planting into a one-person job. You will find it much more efficient to have someone to fill in the hole whilst you hold the plant in the desired position.

It is essential that the union between rootstock and grafted stem is kept well above ground level. The best plan is to aim to have the old soil mark at the top of the planting hole.

Fork over the bottom of the planting hole and form a low mound at the base. A stake is now essential if a free-standing tree (see page 12) is to be grown – supported types have their own form of support. The stake should remain in position for 4–5 years – with M27, M9, Quince C, Pixy, VVA1 and Gisela 5 rootstocks the stake should remain as a permanent feature through the life of the tree. Do not use the soil removed from the hole for planting the tree – use a planting mixture instead.

### Planting Mixture

Make up the planting mixture in a wheelbarrow on a day when the soil is reasonably dry and friable – 1 part topsoil, 1 part well-rotted organic matter and 3 handfuls of bone meal per barrow load. Keep this mixture in a shed or garage until you are ready to start planting.

It is just as important to use a planting mixture for filling the hole when planting a container-grown tree. Roots hate to move from loamless compost into a mineral soil which is practically devoid of organic matter. The recommended planting mixture provides a suitable bridge between the two.

# PLANTING A BARE-ROOTED TREE

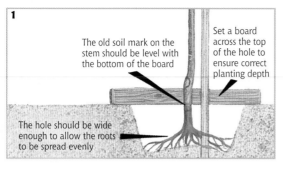

**1**

The old soil mark on the stem should be level with the bottom of the board

Set a board across the top of the hole to ensure correct planting depth

The hole should be wide enough to allow the roots to be spread evenly

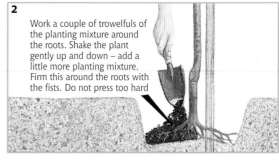

**2**

Work a couple of trowelfuls of the planting mixture around the roots. Shake the plant gently up and down – add a little more planting mixture. Firm this around the roots with the fists. Do not press too hard

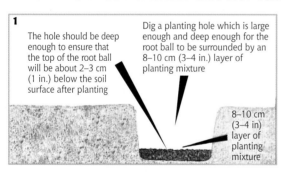

**3** Half-fill the hole with more planting mixture and firm it down. Depending on the size of the plant, do this by gentle treading or by pressing with the fists. On no account should you tread heavily. Start firming at the outer edge of the planting hole, working gradually towards the centre

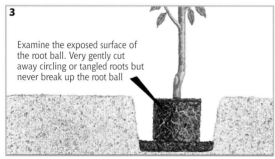

**4**

Add more planting mixture until the hole is full. Firm once again and then loosen the surface. Spread a little soil around the stem to form a low dome

When planting is finished, build a shallow ring of soil. This will form a water-retaining basin

# PLANTING A CONTAINER-GROWN TREE

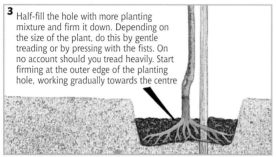

**1**

The hole should be deep enough to ensure that the top of the root ball will be about 2–3 cm (1 in.) below the soil surface after planting

Dig a planting hole which is large enough and deep enough for the root ball to be surrounded by an 8–10 cm (3–4 in.) layer of planting mixture

8–10 cm (3–4 in) layer of planting mixture

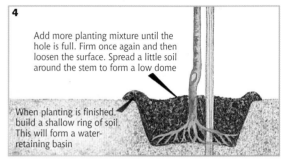

**2**

Water the container thoroughly before planting

Cut down the side of the container when it is stood on the base of the hole. Remove very carefully

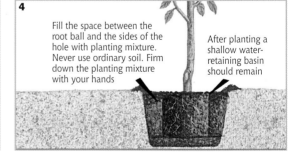

**3**

Examine the exposed surface of the root ball. Very gently cut away circling or tangled roots but never break up the root ball

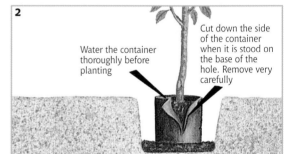

**4**

Fill the space between the root ball and the sides of the hole with planting mixture. Never use ordinary soil. Firm down the planting mixture with your hands

After planting a shallow water-retaining basin should remain

# AFTER PLANTING

Proper after-care is just as important as good planting. Once the new tree is in place it should be watered in thoroughly. In spring place a mulch of rotted manure, garden compost, bark, or black polythene sheeting around the tree – this mulch should extend about 45 cm (18 in.) from the stem.

During the first growing season keep the soil moist and occasionally spray the leaves on warm evenings. Remove all blossom in the first spring after planting.

Check the ties at least once a year – make sure the trunk is not being strangled.

# APPLES

We grow more apples both commercially and in the garden than any other fruit tree – the apple, unlike the pear, is completely at home in this country. There are four basic types, but the first two are of little interest to the home fruit grower. The cider apple with its massive yields (up to 450 kg/1,000 lb per tree) has no place on our plot, although the crab apple with its colourful fruits is grown as a decorative tree and is sometimes used for making jelly.

The choice for the gardener is between the dessert (eating) varieties and the culinary (cooking) varieties, although just a few are dual-purpose. The variety you pick can be bought in a number of growth forms (see page 12) and the rootstock used will largely determine the eventual height (see page 13). A mature dwarf bush will grow about 1.8 m (6 ft) high and provide about 18 kg (40 lb) of fruit; a mature standard will reach 6–7.6 m (20–25 ft) and give yields of 90–180 kg (200–400 lb).

The usual advice is to grow a bush if space is reasonably plentiful. If space is limited then a dwarf bush or a row of cordons is a much better choice. An espalier is the favoured supported form to clothe a wall or fence. We may have to add to this standard advice – the introduction of the compact columnar tree (page 12) is a new trouble-free form for the small garden and the step-over tree (page 12) is a recent introduction which casts little shade in the tiny plot.

Whichever form you choose, plant correctly in properly prepared soil. Finding a suitable site rarely offers a problem, but even the long-suffering apple has its pet hates. Salt-laden sea air creates a problem and so does a shallow alkaline soil. Culinary varieties are more successful than dessert types when things are not quite right, such as rainfall over 1 m (40 in.) per annum, altitude over 142 m (500 ft), partial shade, nitrogen-rich soil or partly impeded drainage.

Plant at the recommended distance (page 23) even though it will seem that the trees are far too widely spaced at first. Follow the rules of correct care set out in this section and even a small number of trees can provide you with fruit from August right through to May.

Many gardeners make do with the old trees they acquired with the house. You can undertake a rejuvenation programme (see page 26) which takes about three seasons, but it is often better to start again, choosing a fresh spot in the garden for the new trees.

| | NOV | DEC | JAN | FEB | MAR | APR | MAY | JUN | JUL | AUG | SEP | OCT |
|---|---|---|---|---|---|---|---|---|---|---|---|---|
| **Plant** – bare-rooted trees | | | | | | | | | | | | |
| **Plant** – container-grown trees | | | | | | | | | | | | |
| **Prune** – all trees | | | | | | | | **Prune** – supported types | | | | |
| **Spray** – as required | | | | | | | | | | | | |
| **Pick** | | | | | | | | | | | | |
| **Store** | | | | | | | | | | | | |

# Tree Types

Apples can be trained in various ways – the choice depends on the needs of the gardener. Some books divide the various growth types into 'unrestricted' and 'restricted', but this rather vague concept makes the classification of the dwarf pyramid and spindlebush rather difficult.

It is more straightforward to think of **free-standing** and **supported** types. Free-standing trees are grown in the open without any horizontal supports – maintenance is usually straightforward once the framework has been established. A supported tree is grown against a fence, framework of wire or a wall and is pruned in summer. Such types require more work than a bush and the yield is lower.

## FREE-STANDING TYPES

### Bush

The most popular type. Bushes have an open centre and a short trunk, bushes 60–75 cm (2–2½ ft), dwarf bushes 45–60 cm (1½–2 ft). They soon come into fruit and are easy to maintain. The mature size depends on the rootstock – 2 m (6 ft) high on M27 to 6 m (18 ft) on MM106. Not suitable for planting in the lawn.

### Standard

Large trees grown on semi-vigorous or vigorous rootstocks – they are only suitable where there is plenty of space and top yields are required. The standard tree has a 2–2.1 m (6–7 ft) high trunk – the half-standard trunk is 1.2–1.35 m (4–4½ ft). These 5–8 m (15–25 ft) tall trees are difficult to care for.

### Pyramid

A pyramid is similar to a bush but the central leader is maintained to give a broadly conical shape. The dwarf pyramid, up to 2.1 m (7 ft) high with a 1.2 m (4 ft) spread, is a good choice for a tub or small plot, but it requires careful summer pruning.

### Spindlebush

A variation of the pyramid – the basic difference is that the side branches are permanently tied down so that they are kept almost horizontal. Not for the garden – the tall central stake and the surrounding ring of vertical strings make it unsightly.

### Compact column

In recent years, as the average garden has become smaller, different tree shapes and forms have been introduced for growing fruit in confined spaces. Some, such as the Ballerina and Colonnade trees, form a compact column of a single main stem with hardly any side branches. Other dwarfing forms are used as container plants for patio and terrace gardening. Many of these newer forms need little or no pruning.

## SUPPORTED TYPES (TRAINED TREES)

### Cordon

A single-stemmed tree which is planted at 45° and tied to a permanent support system such as a fence. A dwarfing rootstock is generally used and vigorous varieties are usually avoided. An excellent way of growing several varieties in a restricted space.

### Espalier

The pairs of branches are stretched horizontally to form a series of tiers at 45 cm (18 in.) intervals. It is more decorative than the cordon, but it takes up more space and it is more difficult to maintain. Buy trees which have already been trained.

### Fan

An attractive form when planted against a wall, but it needs space – a height of 2 m (6 ft) and a spread of about 3 m (10 ft). Careful training is essential. Not a popular tree form for apples – the fan is much more widely used for cherries, plums and peaches

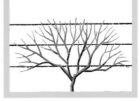

### Step-over

An old idea which attracted interest when re-introduced in 1986. A single-tier espalier is grown along a wire support stretched about 30 cm (12 in.) above ground level. Extremely dwarfing rootstock is used, and the result is an apple edging for bed or border.

# Rootstocks

Apples generally grow badly on their own roots. This has been known for hundreds of years and grafting desirable varieties on to rootstocks has been practised for centuries. But it is only since the end of World War II that the dwarfing rootstock has been widely introduced to the home gardener, and this has made tree-fruit growing possible in small gardens.

The rootstocks we use these days were developed at East Malling Research Station (the M series) and Merton (the MM series). The use of a dwarfing stock brings about all sorts of advantages. Trees come into fruit more quickly and the apples are usually more brightly coloured. Picking becomes a simple task and pruning is much less complicated. Spraying is possible with simple equipment.

It would seem that dwarfing stock should be an automatic choice, but this is not so. Very dwarfing stock needs fertile soil, regular watering and yields are much lower. Thus, it is necessary to choose semi-dwarfing stock or even vigorous stock for poor soil planting.

## M27
### Extremely dwarfing

A newer rootstock which after 10–15 years produces a bush which is only 2 m (6 ft high). Obviously a good choice for container growing, and for a dwarf bush or cordon of a vigorous variety. It is the rootstock used for step-over trees – see page 12. Permanent staking is necessary – the branches may have to be supported. Good soil and regular feeding, watering and weeding are essential.

## M9
### Very dwarfing

An established dwarfing rootstock, producing a bush which is about 3 m (10 ft) across when mature. It will grow under average soil conditions but needs good soil to thrive. It is popular for dwarf bushes, dwarf pyramids, cordons and small espaliers. M9 is a good choice for a small garden with fairly good soil. Permanent staking is necessary, and so is routine feeding and watering.

## The effect of the rootstock

| Rootstock | Approximate height of mature tree | Approximate yield | Age of tree when first fruits appear | Age of tree when full cropping capacity is reached |
|---|---|---|---|---|
| M27 | 1.5–1.8 m (5–6 ft) | 7–11 kg (15–25 lb) | 2–3 years | 4–5 years |
| M9 | 2.4–3 m (8–10 ft) | 16–20 kg (35–45 lb) | 3–4 years | 5–6 years |
| M26 | 3–3.6 m (10–12 ft) | 29–34 kg (65–75 lb) | 3–4 years | 5–6 years |
| MM106 | 4.2–5.4 m (14–18 ft) | 40–50 kg (90–110 lb) | 4–5 years | 7–8 years |
| MM111 & M2 | 5.4–7.6 m (18–25 ft) | 72–162 kg (160–360 lb) | 6–7 years | 6–9 years |

## M26
### Dwarfing

A rootstock which produces an effect which is quite similar to M9. It is rather more vigorous and a little stronger all round. The bush tends to be about 60 cm (2 ft) taller and wider and staking is only needed for about 5 years. This is a good stock for average soil conditions where compact growth is required – the crop is appreciably greater than on an M9 tree.

## MM106
### Semi-dwarfing

Called semi-vigorous in some catalogues – it is the most widely used of all rootstocks. The unlabelled tree at the garden centre is probably on MM106 – it is the best choice for bushes to be grown in the average garden under average conditions. MM106 is nearly always used for the weaker varieties where more dwarfing stocks would produce under-sized bushes. This rootstock can be used for half-standards but it is rather too vigorous for cordons unless the soil is poor.

## MM111 and M2
### Vigorous

There is no place for these rootstocks in most gardens – the resulting tree is far too large and wide-spreading. Trees come into fruit very slowly, and on good soil you will need ladders and extra tools for pruning, spraying and picking. However, they have a number of special uses. MM111 is the right choice for a standard to be grown in a large garden or for a medium-sized bush to be grown on poor soil such as an infertile loamy sand.

# Varieties A–Z

There are hundreds of varieties to choose from. Some are available at almost every nursery and others are restricted to just one or two specialist catalogues. Listed here are 56 of them – the ones you are most likely to find in the garden centre or mail order list. The apples described are certainly not a 'recommended' or 'best-buy' selection – a few of them are a poor choice.

Read the descriptions carefully to see if the variety is right for you – there is no such thing as the 'perfect' apple. Avoid very vigorous types such as *Bramley's Seedling* if space is limited. Choose an early or mid-season fruit variety (see below) if you live in a northern district and reject a canker-prone type if your ground stays cold and wet in spring. Frost is an important consideration. If your garden seems more prone than most to suffer from frost then you should choose a mid-season/late or late-flowering variety (see 'Pollination Groups'). Alternatively, look through the A–Z list for an apple with a good frost resistance – do avoid *Cox's Orange Pippin* and *Bramley's Seedling*. When choosing a culinary variety, think of the end-product you require – *Rev W. Wilks* cooks down to a golden fluff whereas *Lane's Prince Albert* keeps its shape.

Pick your apples in peak condition – commercial fruit is often picked too early. Remember the general rule – home-grown 'English' varieties (*Cox's Orange Pippin, Worcester Pearmain*, etc.) taste better than shop-bought ones, but home-grown 'foreign' varieties (*Granny Smith, Sturmer Pippin*, etc.) taste worse.

You will find some odd words in the descriptions – look them up in the glossary as they may affect your decision. Biennial bearing means that you can have a very disappointing crop every other year. Tip-bearing means that most of the fruit is borne at the end instead of along the length of the branch which is the much more usual spur-bearing habit. The months listed in the A–Z descriptions refer to southern England – add on a couple of weeks for the north. For apples which need to be stored note that the start of the period of use may be up to 2 months after the recommended time for picking.

| Fruit size | Diameter of fruit |
|---|---|
| Very large | over 8 cm (3½ in.) |
| Large | 7–8 cm (2¾–3½ in.) |
| Medium | 5–7 cm (2–2¾ in.) |
| Small | less than 5 cm (2 in.) |

| Fruiting season | Picking time | Storage period |
|---|---|---|
| Early | July–early September | Nil – eat within 7 days |
| Mid-season | September–October | Limited – 2–3 weeks |
| Late | October–November | Prolonged – 1–6 months depending upon variety |

## POLLINATION GROUPS

A few apples are self-fertile, capable of setting some fruit with their own pollen – examples include *James Grieve* and *Arthur Turner*. In practically all cases, however, it is necessary to have a pollination partner nearby – another variety which flowers at approximately the same time, enabling cross-pollination to take place.

In urban areas where apples occur in many gardens there is often no need to provide a pollination partner in your own garden – there are usually enough suitable ones in surrounding plots. But if apples are not common in your area then it will be necessary to plant a suitable pollen provider in your garden.

Apples are classified into 4 flowering season groups – see below. A pollination partner is a variety which is ideally in the same group or else in the one immediately above or just below the group which includes the apple in question.

The so-called triploid varieties pose a special problem. Here the variety is an extremely poor pollinator, so non-triploid varieties need to be grown nearby to act as pollination partners – these pollinate both the triploid variety and each other.

| | | |
|---|---|---|
| **A** Early flowering | Beauty of Bath Egremont Russet George Cave Irish Peach Lord Lambourne | Rev W. Wilks St Edmund's Pippin |
| | | **AT** Triploid – needs two partners Ribston Pippin |
| **B** Mid-season flowering | Arthur Turner Bountiful Charles Ross Cox's Orange Pippin Discovery Early Victoria Elstar Epicure Fiesta Fortune Greensleeves Grenadier James Grieve Katy Kidd's Orange Red | Lane's Prince Albert Meridian Redsleeves Red Windsor Scrumptious Spartan Sunset Worcester Pearmain |
| | | **BT** Triploid – needs two partners Blenheim Orange Bramley's Seedling Jupiter |
| **C** Mid-season/ late flowering | Annie Elizabeth Ashmead's Kernel Christmas Pippin Cox's Self-fertile Ellison's Orange Gala Golden Delicious Golden Noble Herefordshire Russet Howgate Wonder Laxton's Superb | Lord Derby Orleans Reinette Peasgood Nonsuch Pixie Tickled Pink Tydeman's Late Orange Winston |
| | | **CT** Triploid – needs two partners Jonagold |
| **D** Late flowering | Edward VII Newton Wonder | **DT** Triploid – needs two partners Suntan |

## FAMILY TREES

The need for a pollination partner can create a difficulty in a small garden where there is room for only one tree. Growing a family tree can be the answer. Here 2–4 different but compatible varieties are grafted on to one tree – ensuring cross-pollination plus a prolonged cropping period. A choice of rootstocks and an extensive range of varieties are available in this form.

Typical offerings include:

*Charles Ross/Grenadier/Worcester Pearmain*
*Discovery/Fortune/Sunset*
*Cox's Orange Pippin/James Grieve/Spartan*

Avoid combinations which include *Bramley's Seedling* or *Golden Delicious* as these varieties tend to take over.

### Annie Elizabeth

**Type:** Culinary

**Fruit size:** Large

**Skin colour:** Golden, flushed, speckled and striped with pinkish-red

**Pollination group:** C

**Picking time:** Mid-October

**Period of use:** December–May

No cooking apple has better keeping qualities than old Annie – it lasts for months in store. An upright tree which spurs freely and is suitable for northern climes. The flesh is acid and crisp – texture is rather dry. *Annie Elizabeth* has been around for a long time and was once popular with commercial growers, but it is no longer in the top-seller lists.

Annie Elizabeth

### Arthur Turner AGM

**Type:** Culinary

**Fruit size:** Large

**Skin colour:** Greenish-yellow, flushed orange-brown

**Pollination group:** Partly self-fertile: B

**Picking time:** August–September

**Period of use:** August–October

*Arthur Turner* is not recommended by all the experts, but it certainly has its share of plus points. The pink blossom is outstanding – no apple provides a finer display. Growth is upright and this variety is reliable in northern districts. Crops are heavy and the rather dry mixture of the fruit makes it an excellent baking apple. Quite freely available.

Arthur Turner

### Ashmead's Kernel AGM

**Type:** Dessert

**Fruit size:** Medium

**Skin colour:** Yellowish-green, mostly covered with mid-brown russet

**Pollination group:** C

**Picking time:** Mid-October

**Period of use:** December–March

Nearly 300 years old and still regarded as one of the best late dessert apples. The yellow flesh is crisp and the flavour is rated as superb. Scab resistance is good and the fruit has excellent keeping qualities. Quality is not matched by quantity – it is a light and erratic cropper which cannot be recommended for exposed northern gardens.

Beauty of Bath

### Beauty of Bath

**Type:** Dessert

**Fruit size:** Small–medium

**Skin colour:** Pale yellow, flushed and speckled with bright red

**Pollination group:** A

**Picking time:** Early August

**Period of use:** Early August

Once a very popular early dessert apple, but commercial growers now choose better varieties. The tree has a spreading habit – the taste is sweet but somewhat tart and nothing special. You will find it in numerous catalogues but it is no longer recommended by the experts. Premature fruit drop is a problem and cropping is slow to start.

### Blenheim Orange AGM

**Type:** Dessert/culinary

**Fruit size:** Large

**Skin colour:** Golden, flushed and striped with dull red and fine brown russet

**Pollination group:** BT

**Picking time:** Early October

**Period of use:** November–January

After nearly two centuries it is still regarded as the best of all dual-purpose apples. The creamy flesh is crisp and dry with an excellent nutty flavour. Mildew resistance is good and yields are heavy. Growth is very vigorous – make sure it is on dwarfing stock. Succeeds in the north. It is a biennial bearer and scab can be a problem.

Blenheim Orange

### Bountiful

**Type:** Culinary

**Fruit size:** Large

**Skin colour:** Pale green, striped with orange-red

**Pollination group:** B

**Picking time:** Late September

**Period of use:** September–January

The first new cooking apple for decades – a good choice for people who want large fruit but haven't the space for a *Bramley's Seedling*. This compact variety pollinates freely and does not suffer from mildew. Heavy crops are produced and *Bountiful* can be bought as a cordon. Fruit is quite sweet for a cooker – extra sugar is not required.

Bountiful

### Bramley's Seedling AGM

**Type:** Culinary

**Fruit size:** Very large

**Skin colour:** Yellowish-green, lightly striped with red

**Pollination group:** BT

**Picking time:** Mid-October

**Period of use:** November–March

By far the most popular cooking apple. The reason is easy to see – heavy crops of extra-large fruit with creamy flesh which is juicy and full of flavour. But think before making it your first choice – it is too vigorous for small gardens and it is often a biennial bearer. Neither scab nor frost resistance is good. A partial tip-bearer.

**Charles Ross**

**Cox's Orange Pippin**

**Discovery**

**Early Victoria**

**Edward VII**

### Charles Ross AGM

Some apple varieties are chosen for their outstanding flavour – this one's claim to fame is the beauty of the fruit. Large, round, colourful and showy – a popular sight on the show bench. The flesh is sweet and juicy at first but becomes dry and flavourless with long storage. This eater/baker is widely available and does well in limy soils.

**Type:** Dessert/culinary

**Fruit size:** Large

**Skin colour:** Yellowish-green, flushed and striped with orange-red

**Pollination group:** B

**Picking time:** Mid-September

**Period of use:** October–December

### Christmas Pippin AGM

A recent introduction, discovered as a seedling growing close to a motorway. It has a creamy flesh and aromatic flavour with a high natural sugar content, but with some sharpness too. Similar to Cox's Orange Pippin, but much easier to grow, with reliable heavy crops nationwide. Some fruit thinning is usually needed. Suitable for colder areas and is more resistant to frost.

**Type:** Dessert

**Fruit size:** Medium

**Skin colour:** Pale yellow, flushed red

**Pollination group:** C

**Picking time:** October

**Period of use:** November–January

### Cox's Orange Pippin

Regarded by most people as the best of all eating apples – aromatic, juicy, crisp, superb flavour, etc. Still, it is in most cases one for your weekly shopping list rather than the garden. It is not suitable for northern districts and even in the south it is temperamental and susceptible to frost and disease. It needs good soil and knowledgeable attention.

**Type:** Dessert

**Fruit size:** Medium

**Skin colour:** Golden, flushed with orange-red. Patches of brown russet

**Pollination group:** B

**Picking time:** October

**Period of use:** October–January

### Cox's Self-fertile

A Cox form which is a better grower and more reliable cropper. A much better choice than the standard Cox, it retains the excellent aromatic flavour and crisp, crunchy texture. A moderately vigorous variety, with an upright habit spreading with age, the tree has better resistance to disease than the original Cox, but requires a well-drained, sheltered site to do well.

**Type:** Dessert

**Fruit size:** Medium

**Skin colour:** Golden, flushed with orange-red. Patches of brown russet

**Pollination group:** Self-fertile: C

**Picking time:** October

**Period of use:** October–January

### Discovery AGM

An early variety with firmer flesh and better keeping qualities than most others which ripen in August–September. It is earlier and better than its parent *Worcester Pearmain*. Disease resistance and frost tolerance are good, and the moderate growth habit makes it suitable for the smaller garden. It can be slow to start cropping – a spur and tip-bearer.

**Type:** Dessert

**Fruit size:** Small–medium

**Skin colour:** Yellow, almost entirely covered with bright crimson

**Pollination group:** B

**Picking time:** August

**Period of use:** August–September

### Early Victoria AGM

*Early Victoria* (also known as *Emneth Early*) is the first cooker of the season. Fruits are ready for picking in late July and the greenish-white flesh cooks to a sweet fluff. Trees are very hardy and yields are high. Growth is upright and compact – the major problem is that thinning of the trusses is essential or biennial bearing and small fruits will result.

**Type:** Culinary

**Fruit size:** Medium

**Skin colour:** Yellowish-green

**Pollination group:** B

**Picking time:** July

**Period of use:** July–August

### Edward VII AGM

Late to flower and late to fruit – this hardy variety is a good choice for northern areas or gardens with a frost pocket. Growth is compact and upright, so it is on the recommended list for small gardens. Scab resistance is good. The acid flesh turns pink after cooking. *Edward VII* is only a moderate cropper, and is often slow to start bearing fruit.

**Type:** Culinary

**Fruit size:** Large

**Skin colour:** Yellowish-green

**Pollination group:** D

**Picking time:** Mid-October

**Period of use:** December–April

**Egremont Russet AGM**

You will find this one in nearly all the catalogues. It has been around for a hundred years but it's still one of the best russets you can buy. The skin is rough and the flesh is crisp. Flavour is very good – nutty and sweet. Growth is upright and the fruit is rather small. An advantage is the frost tolerance of its blossom, but bitter pit can be a problem.

**Type:** Dessert

**Fruit size:** Small–medium

**Skin colour:** Golden, large patches of pale brown russet

**Pollination group:** A

**Picking time:** Late September

**Period of use:** October–December

---

**Ellison's Orange AGM**

A hardy apple which is easy to grow. It is praised to the skies in some catalogues, and is sometimes listed as *Red Ellison*. Scab and frost resistance are good, but it is still not really a good choice – the fruit is very juicy but the rather odd aniseed flavour is not enjoyed by everybody. Canker can be a problem, and so can biennial bearing.

**Type:** Dessert

**Fruit size:** Medium

**Skin colour:** Greenish-yellow, flushed and striped with red

**Pollination group:** C

**Picking time:** Mid-September

**Period of use:** September–October

---

**Elstar AGM**

A newer variety from Holland which is not yet in the textbooks and is restricted to the catalogues of specialist fruit growers. Elstar is expected to become a popular variety – it crops heavily, stores well and both juiciness and flavour are outstanding. Thin fruitlets to avoid biennial bearing – not a tree for beginners.

**Type:** Dessert

**Fruit size:** Medium

**Skin colour:** Greenish-yellow, flushed with bright red

**Pollination group:** B

**Picking time:** October

**Period of use:** October–January

---

**Epicure AGM**

You will find it in numerous catalogues – often listed as *Laxton's Epicure*. A good choice for beginners and cold districts – it is both hardy and easy to grow. The fruit is prettily striped and juicy, but it tends to be small and the flavour rapidly deteriorates after picking. A compact tree with few branches.

**Type:** Dessert

**Fruit size:** Small

**Skin colour:** Yellow, striped with red

**Pollination group:** B

**Picking time:** Mid-August

**Period of use:** Mid-August–September

---

**Fiesta AGM**

An outstanding new variety for people who would like to grow *Cox's Orange Pippin* but can't. Cox is one of the parents and all of the aroma and nuttiness are present. The flavour, however, is not quite as good, despite the claim in a few catalogues. The main points are that *Fiesta* is reliable and it crops heavily. Growth is rather pendulous.

**Type:** Dessert

**Fruit size:** Medium–large

**Skin colour:** Yellow, flushed and striped with bright red

**Pollination group:** B

**Picking time:** Early October

**Period of use:** October–March

---

**Fortune AGM**

*Fortune* (*Laxton's Fortune*) is a popular variety which has a place in most catalogues. It appeals because it has a compact growth habit, is easy to grow and the fruit is sweet and juicy. However, it's not a winner these days – problems include canker, biennial bearing and a loss of crispness soon after picking.

**Type:** Dessert

**Fruit size:** Medium

**Skin colour:** Golden-yellow, flushed and striped with red

**Pollination group:** B

**Picking time:** Early September

**Period of use:** September–October

---

**Gala AGM**

Not widely available, although you will find it in some of the specialist catalogues. *Gala* is sometimes praised for its shiny red skin and crisp, sweet flesh; the degree of redness depends on the amount of sunshine. It crops heavily but you can do better. *Gala* is prone to both scab and canker and the fruit can be quite small if not thinned.

**Type:** Dessert

**Fruit size:** Medium

**Skin colour:** Yellow, heavily flushed with red

**Pollination group:** C

**Picking time:** October

**Period of use:** October–January

George
Cave

Greensleeves

Grenadier

Herefordshire
Russet

Howgate
Wonder

### George Cave

Regarded by some as the best of the early varieties. *George Cave* crops heavily and regularly, the first fruits being ready in late July in favourable districts. Do not wait too long before picking – early fruit drop can occur. The fruit is described as refreshing – soft, juicy and acidic, but with no special merit. A tip- and spur-bearer.

**Type:** Dessert
**Fruit size:** Small–medium
**Skin colour:** Green, flushed and striped with red
**Pollination group:** A
**Picking time:** August
**Period of use:** August

### Golden Delicious AGM

Don't be tempted to plant this one because you like the shop-bought fruit, with its characteristic ribs and the bumps at the crown. *Golden Delicious* needs sunnier climes than Britain to give of its best – flavour and size will disappoint after a poor summer. Yields are usually heavy and the fruit stores well. A good pollinator for other varieties.

**Type:** Dessert
**Fruit size:** Medium
**Skin colour:** Greenish, or golden-yellow
**Pollination group:** C
**Picking time:** Late October
**Period of use:** November–February

### Golden Noble AGM

Highly praised in many textbooks as one of the best cookers. The flesh becomes a golden froth with a delicious flavour, but you will have to turn to a specialist fruit grower if you want to buy one. It succeeds in cold districts and crops regularly but it is susceptible to scab and canker. A tip- and spur-bearer.

**Type:** Culinary
**Fruit size:** Medium–large
**Skin colour:** Yellow. Pale russet may be present
**Pollination group:** C
**Picking time:** Early October
**Period of use:** October–December

### Greensleeves AGM

An outstanding modern variety – this is the one to grow if you are a *Golden Delicious* fan. The tree is hardy, reliable, self-fertile and easy to manage. Fruit appears early in the life of the tree and very heavy crops can be expected. Unfortunately the refreshing flavour soon fades in store. A tip- and spur-bearer.

**Type:** Dessert
**Fruit size:** Medium
**Skin colour:** Pale green turning to pale yellow
**Pollination group:** B
**Picking time:** September
**Period of use:** September–November

### Grenadier AGM

*Grenadier* is the standard early cooking apple seen on the greengrocers' shelves – green, flattish and distinctly ribbed. It is also a popular garden variety as it is hardy, crops heavily and has a compact growth habit. Scab resistance is good and it cooks down to a white fluff with an excellent flavour. Poor storage quality is the major problem.

**Type:** Culinary
**Fruit size:** Medium–large
**Skin colour:** Yellowish-green
**Pollination group:** B
**Picking time:** Mid-August
**Period of use:** August–September

### Herefordshire Russet AGM

The quality of this new variety is exceptional. The tree is largely disease-free with heavy crops of evenly sized, large fruits, frost tolerance and a long picking and eating season. The white-fleshed fruits have a rich, aromatic and nutty flavour and a crisply satisfying texture. With good frost tolerance, this apple can be grown successfully throughout the UK.

**Type:** Dessert
**Fruit size:** Medium–large
**Skin colour:** Golden brown patches of pale russet
**Pollination group:** Self-fertile: C
**Picking time:** October
**Period of use:** October–January

### Howgate Wonder

A good choice if you live in the north or in a frosty area and if you don't mind a vigorous, spreading tree in your garden. Crops are very heavy and the extra-large, striped fruit is suitable for the autumn horticultural show. Unfortunately the flavour is not as good as many of the catalogues indicate – it is often insipid.

**Type:** Culinary
**Fruit size:** Very large
**Skin colour:** Greenish-yellow, flushed and striped with pale red
**Pollination group:** C
**Picking time:** October
**Period of use:** November–March

**Type:** Dessert

**Fruit size:** Small–medium

**Skin colour:** Yellow, flushed and striped with orange and red

**Pollination group:** A

**Picking time:** Late August

**Period of use:** August–September

---

**Type:** Dessert

**Fruit size:** Medium–large

**Skin colour:** Yellow, speckled and striped with orange

**Pollination group:** Early self-fertile: B

**Picking time:** Early September

**Period of use:** September–October

---

**Type:** Dessert

**Fruit size:** Large–very large

**Skin colour:** Greenish-yellow, lightly flushed and mottled with red

**Pollination group:** CT

**Picking time:** Mid-October

**Period of use:** November–March

---

**Type:** Dessert

**Fruit size:** Medium

**Skin colour:** Golden-yellow, mostly flushed and striped with red

**Pollination group:** BT

**Picking time:** Mid-September

**Period of use:** October–March

---

**Type:** Dessert

**Fruit size:** Small–medium

**Skin colour:** Greenish-yellow, heavily flushed with bright red

**Pollination group:** B

**Picking time:** Early September

**Period of use:** September–October

---

**Type:** Dessert

**Fruit size:** Small–medium

**Skin colour:** Yellow, heavily flushed with red. Patches of pale brown russet

**Pollination group:** B

**Picking time:** Early October

**Period of use:** November–February

---

**Type:** Culinary

**Fruit size:** Large

**Skin colour:** Grass-green, striped with dull red

**Pollination group:** B

**Picking time:** Early October

**Period of use:** November–March

## Irish Peach

You will find this one in a number of guides to recommended varieties, but not in many catalogues. This recommendation is based on the outstanding flavour and juiciness of fruit picked straight from the tree. The lack of popularity is based on its reluctance to reach the fruiting stage, the poor quality of stored fruit and its tip-bearing nature.

## James Grieve AGM

An excellent choice, especially for a difficult site, if you want a reliable and hardy early variety. It crops heavily and regularly, and the soft-textured fruit is both juicy and tangy. Thinning is often necessary and the keeping quality is rather poor. Does better in the north than in western areas, where canker may be a problem.

## Jonagold AGM

A modern apple from the US with *Golden Delicious* and *Jonathan* as its parents. Fruit quality is excellent – crisp, juicy and full of flavour. The colour is sometimes insipid, and its vigour and spreading habit may be a problem in a small garden. Yields are very high and young trees quickly come into fruit.

## Jupiter AGM

Introduced in 1982 as the variety which had all the flavour of *Cox's Orange Pippin* and none of its unreliability. Cropping is very heavy and growth is vigorous – make sure that it is grafted on dwarfing stock. It is quite easy to find a supplier, but its Cox-substitute role is being taken over by others like *Sunset* and *Fiesta*.

## Katy

*Katy*, *Katja* or *Katya*. This early variety came to us from Sweden – not surprisingly, it is a good choice for northern gardens. It is similar in many ways to its parent *Worcester Pearmain*, but *Katy* is generally considered to be a better choice these days. Cropping is heavy and regular – fruit is juicy and refreshing.

## Kidd's Orange Red AGM

This New Zealand variety was expected to become popular because it is similar to *Cox's Orange Pippin* in flavour and texture but has none of its cultural problems. It is a heavy and regular cropper and this variety spurs freely. It is suitable for northern gardens and has quite a good resistance to disease. However, it never became a best-seller.

## Lane's Prince Albert AGM

One of the small group of popular cooking apples which you will find in most catalogues. It has many good points. The tree is hardy, compact and spurs freely. The fruit keeps better than *Bramley's Seedling* and does not break up when cooked. Still, it has its problems. It is susceptible to mildew and has too little vigour for poor soil.

James Grieve

Jonagold

Jupiter

Katy

Kidd's Orange Red

**Laxton's Superb**

**Lord Derby**

**Newton Wonder**

**Orleans Reinette**

### Laxton's Superb

Perhaps the most widely grown of all late dessert varieties. There are some good points – the flowers have above-average frost resistance and the fruit stores well. But you can do better these days. It is a biennial bearer, susceptible to scab and a little too vigorous for the small garden. Thinning is usually necessary.

**Type:** Dessert

**Fruit size:** Medium–large

**Skin colour:** Greenish-yellow, flushed with dark red. Patches of russet

**Pollination group:** C

**Picking time:** Early October

**Period of use:** November–February

### Lord Derby

This old apple is still quite widely available but no longer appears in the lists of recommended varieties. It does not have the keeping qualities you would expect from a cooker picked in September or October, but the variety has several good points. It succeeds in cold and wet areas and has good disease resistance. Flesh turns pink when cooked.

**Type:** Culinary

**Fruit size:** Large

**Skin colour:** Green

**Pollination group:** C

**Picking time:** Late September

**Period of use:** October–December

### Lord Lambourne AGM

This mid-season variety is highly recommended by many experts. The reason is that it can be relied upon to crop both heavily and regularly almost anywhere. Its compact growth habit makes it suitable for small gardens and the brightly coloured fruit is sweet, juicy and aromatic. A tip- and spur-bearer which spurs very freely.

**Type:** Dessert

**Fruit size:** Medium

**Skin colour:** Greenish-yellow, flushed and striped with red

**Pollination group:** A

**Picking time:** Mid-September

**Period of use:** October–November

### Meridian

A newer variety from East Malling Research Station, this is an excellent choice for the garden. Very good flavour, with a crisp, crunchy texture and plenty of sweetness but a good balance of sharp acidity as well. This variety stores well and will appeal to most tastes. Has good disease resistance, making it popular with organic gardeners.

**Type:** Dessert

**Fruit size:** Medium–large

**Skin colour:** Pale green, striped orange

**Pollination group:** B

**Picking time:** Mid-September

**Period of use:** October–January

### Newton Wonder AGM

This very vigorous apple is not really suitable for small gardens. It is recommended for its heavy yields of fine-flavoured fruit, but it is a biennial bearer and prone to bitter pit. The fruit is juicy, acid and boils down to a golden fluff. Not really a dual-purpose variety, but stored fruit can be used as a dessert apple.

**Type:** Culinary

**Fruit size:** Very large

**Skin colour:** Golden-yellow, flushed and striped with red

**Pollination group:** D

**Picking time:** Mid-October

**Period of use:** November–December

### Orleans Reinette

An 18th-century variety still offered for sale because of its excellent fruit quality – crisp flesh and a delicious nutty flavour. The catalogues, however, tend to make light of its problems. Yields are rather low and irregular – it is a biennial bearer and is prone to scab in wet soils. This one does not store well – fruit often shrivels in store.

**Type:** Dessert

**Fruit size:** Medium–large

**Skin colour:** Yellow, flushed orange-red. Patches of grey-brown russet

**Pollination group:** C

**Picking time:** Mid-October

**Period of use:** November–January

### Pixie AGM

You will have to search for this one, which is a shame. The fruit quality is exceptional – attractive, crisp, aromatic, juicy, refreshing etc. The problem is that the fruits are rather small, but this should not really discourage gardeners. The growth properties of the variety are good – cropping is prolific and disease resistance is high. It must be thinned.

**Type:** Dessert

**Fruit size:** Small–medium

**Skin colour:** Yellowish-green, flushed and striped with red

**Pollination group:** C

**Picking time:** Mid-October

**Period of use:** December–March

**Type:** Culinary/dessert

**Fruit size:** Very large

**Skin colour:** Golden, flushed red

**Pollination group:** C

**Picking time:** September

**Period of use:** September–December

### Peasgood Nonsuch

A useful garden variety, which is partially self-fertile and will produce some fruit without a pollinator. Very good flavour, slightly acid, but still sweet and strongly aromatic. Large, handsome, regular-shaped fruits often grown for exhibition. Golden, flushed with red stripes. A Victorian English cooking apple, it makes a sweet, delicately flavoured purée and is superb for baking.

---

**Type:** Dessert

**Fruit size:** Medium

**Skin colour:** Yellowish-green, almost entirely flushed with bright red

**Pollination group:** Partly self-fertile: B

**Picking time:** Late August

**Period of use:** August–September

### Redsleeves

Will *Redsleeves* be a leading early variety in the future? Only time will tell, but the reports from East Malling Research Station trial look promising. Growth is compact, and the fruit is crisp and juicy. Disease resistance is claimed to be excellent and it is reported to need very little pruning. Apples can be stored for up to a month.

**Peasgood Nonsuch**

---

**Type:** Dessert/juice

**Fruit size:** Medium–large

**Skin colour:** Red

**Pollination group:** Self-fertile: B

**Picking time:** September

**Period of use:** September–November

### Red Windsor AGM

This is a highly attractive, brightly red-flushed apple, firmly textured with a richly aromatic, but sweet and honeyed flavour. It is a free-flowering tree which sets well, making it a reliable cropper, often producing very high yields. The tree has a naturally compact habit. Disease resistance and frost tolerance are good. Ideal for growing in northerly areas.

**Rev W. Wilks**

---

**Type:** Culinary

**Fruit size:** Very large

**Skin colour:** Pale green, flushed and striped with dark pink

**Pollination group:** A

**Picking time:** Early September

**Period of use:** September–November

### Rev W. Wilks

A compact, short-rooted cooking variety which is specially recommended for small gardens. Disease resistance is good and it is surprising that such a small tree can give high yields of very large fruit. Quality is good – juicy and tangy with a reputation for being an excellent baker. No variety is perfect – *Rev W. Wilks* is a biennial bearer.

---

**Type:** Dessert

**Fruit size:** Medium

**Skin colour:** Yellow, flushed and striped with dull red. Patches of russet

**Pollination group:** AT

**Picking time:** Late September

**Period of use:** October–January

### Ribston Pippin AGM

As you would expect from a parent of *Cox's Orange Pippin*, the russeted fruit of this variety is firm, aromatic and fine-flavoured. *Ribston Pippin* is not widely available as it does have its problems. The tree is prone to canker and the fruit drops easily. It crops regularly but yields are not always heavy. A fruit with a fine past but limited future.

**St Edmund's Pippin**

---

**Type:** Dessert

**Fruit size:** Small–medium

**Skin colour:** Golden, lightly flushed with orange. Patches of brown russet

**Pollination group:** A

**Picking time:** Mid-September

**Period of use:** September–October

### St Edmund's Pippin AGM

Choose this one if you want a russet early in the season. The fruit quality is very high with a distinct tangy sweetness. The texture is firm. The growth habit is compact and cropping is regular. Thinning may be necessary when overcropping takes place. In some years yields are low. A tip- and spur-bearer – unsuitable for growing as a cordon.

---

**Type:** Dessert

**Fruit size:** Medium

**Skin colour:** Bright red

**Pollination group:** Self-fertile: B

**Picking time:** Early September

**Period of use:** September–October

### Scrumptious AGM

An early-season dessert apple and the apples can be picked over a period of several weeks, rather than all at once. Named for its wonderful complexity of flavours, which are sweet, rich and aromatic. An ideal garden variety with frost-resistant blossom and a good pollinator for other varieties. It is a reliable, heavy-cropping tree, with good disease resistance.

**Scrumptious**

Spartan

Sunset

Tydeman's
Late Orange

Worcester
Permain

### Spartan

This is one for the *Red Delicious* or *McIntosh* fan. The smooth-skinned red fruit is white-fleshed, crisp and very juicy. The hardy tree yields moderately well and spurs freely, but canker can be a problem. So can fruit size – thinning is essential if there is an abundance of fruitlets. This variety responds well to feeding.

**Type:** Dessert

**Fruit size:** Medium

**Skin colour:** Yellow, almost entirely flushed with dark red

**Pollination group:** B

**Picking time:** Early October

**Period of use:** November–January

### Sunset AGM

The usual choice where a Cox-like apple is required. This offspring of *Cox's Orange Pippin* has nearly all of the quality of its illustrious parent with hardly any of the bad points. It is hardy, compact and has very good disease resistance. It crops heavily and regularly but the fruit is rather small. With care will keep until January.

**Type:** Dessert

**Fruit size:** Small–medium

**Skin colour:** Golden, flushed and striped with red. Small patches of russet

**Pollination group:** B

**Picking time:** Late September

**Period of use:** October–December

### Suntan AGM

Another Cox-like variety. It scores over *Sunset* by producing larger fruit which can be stored until early spring. It also flowers later so that it escapes frost damage. However, the fruits are more acidic than *Sunset* and they are prone to bitter pit. In addition growth is much more vigorous and two pollinators are needed.

**Type:** Dessert

**Fruit size:** Medium–large

**Skin colour:** Yellowish-green, flushed and striped with red. Patches of russet

**Pollination group:** DT

**Picking time:** Mid-October

**Period of use:** November–February

### Tickled Pink

A very ornamental tree with large crimson-pink flowers in spring and deep purple foliage fading slightly with age. Good-quality, multi-purpose fruits, brightly coloured and attractive, with pink flesh. These can be picked early for cooking or used as a dessert apple in early autumn. The fruits have a crisp texture and mildly aromatic flavour.

**Type:** Dessert/culinary/juice

**Fruit size:** Medium

**Skin colour:** Deep red

**Pollination group:** C

**Picking time:** Late September

**Period of use:** October–November

### Tydeman's Late Orange

The fruit is very similar in texture, colour and taste to its parent *Cox's Orange Pippin*. This variety is easy to grow – spurs are freely produced and the blossom is quite resistant to frost. The storage qualities are excellent – much better than Cox – but yields are only moderate. Thinning is necessary and it is a biennial bearer.

**Type:** Dessert

**Fruit size:** Small–medium

**Skin colour:** Golden, flushed and striped with dull red. Patches of russet

**Pollination group:** C

**Picking time:** Mid-October

**Period of use:** December–April

### Winston AGM

A late dessert apple which is an improvement on *Laxton's Superb*. The flavour and colour are better and so is the resistance to disease. In addition it will set fruit without a pollination partner. Still, this hardy and easy-to-grow variety is not widely listed in the catalogues – perhaps the problem is the need to thin the trusses in June.

**Type:** Dessert

**Fruit size:** Small–medium

**Skin colour:** Yellowish-green, flushed and striped with red

**Pollination group:** Partly self-fertile: C

**Picking time:** Mid-October

**Period of use:** December–April

### Worcester Pearmain AGM

An old favourite – hardy, reliable and resistant to mildew. The trick is to leave the fruits on the tree until they are fully ripe – in that way the flavour and aroma are enhanced. There are problems – it is susceptible to both scab and canker, and it is a tip-bearer. It is slow to start bearing fruit – choose *Discovery* or *Katy* instead.

**Type:** Dessert

**Fruit size:** Small–medium

**Skin colour:** Yellowish-green, almost completely flushed with bright red

**Pollination group:** B

**Picking time:** Early September

**Period of use:** September–October

# Planting

Follow the basic rules for planting set out on pages 9–10. Before putting in the tree the soil should have been limed if it is very acid and some form of windbreak should have been created (see page 33) if the site is exposed.

Remember to plant to the old soil mark – the union with the rootstock should be about 10 cm (4 in.) above ground level. The bulge formed by this union should be on the upper surface when planting cordon apples. Plant cordons in north–south rows with the trees pointing north. Make sure that all stakes and supports are in position before planting – cordons need canes set at 45°.

## Wall support for fans

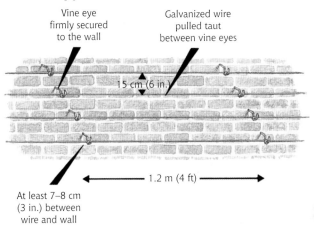

Vine eye firmly secured to the wall

Galvanized wire pulled taut between vine eyes

15 cm (6 in.)

1.2 m (4 ft)

At least 7–8 cm (3 in.) between wire and wall

## Fence support for cordons, espaliers and fans

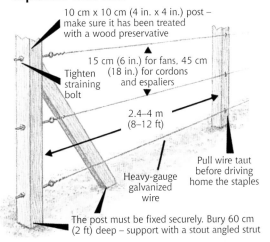

10 cm x 10 cm (4 in. x 4 in.) post – make sure it has been treated with a wood preservative

15 cm (6 in.) for fans, 45 cm (18 in.) for cordons and espaliers

Tighten straining bolt

2.4–4 m (8–12 ft)

Pull wire taut before driving home the staples

Heavy-gauge galvanized wire

The post must be fixed securely. Bury 60 cm (2 ft) deep – support with a stout angled strut

| Tree type | Space between trees | Space between rows | Yield per mature tree |
|---|---|---|---|
| Standard | 9 m (30 ft) | 9 m (30 ft) | 90–180 kg (200–400 lb) |
| Half-standard | 6 m (20 ft) | 6 m (20 ft) | 45–112 kg (100–250 lb) |
| Fan | 4.5 m (15 ft) | – | 7–11 kg (15–25 lb) |
| Bush | 4.5 m (15 ft) | 4.5 m (15 ft) | 27–54 kg (60–120 lb) |
| Espalier | 4.2 m (14 ft) | 1.8 m (6 ft) | 9–14 kg (20–30 lb) |
| Dwarf bush<br>– M27 stock<br>– other stocks | 1.5 m (5 ft)<br>2.7 m (9 ft) | 2.4 m (8 ft)<br>4.5 m (15 ft) | 5.5–7 kg (12–16 lb)<br>18–27 kg (40–60 lb) |
| Spindlebush | 2.1 m (7 ft) | 4.2 m (14 ft) | 27–54 kg (60–120 lb) |
| Dwarf pyramid | 1.5 m (5 ft) | 2.4 m (8 ft) | 4.5–6.5 kg (10–14 lb) |
| Step-over | 1.5 m (5 ft) | 1.8 m (6 ft) | 2.25–2.7 kg (5–6 lb) |
| Cordon | 75 cm (2½ ft) | 1.8 m (6 ft) | 2.25–3 kg (5–7 lb) |
| Compact column | 60 cm (2 ft) | 1.8 m (6 ft) | 4.5–9 kg (10–20 lb) |

*The planting hole should be at least twice the size of the tree's root ball.*

# Feeding and Mulching

A starved tree cannot support a large crop – lack of nutrients can also affect next year's harvest. There are several complex feeding programmes for apples, but a simple and well-tried method is to sprinkle general fertilizer at a rate of 60 g per sq. metre (2 oz per sq. yard) in early spring around the soil under the branches. In April put a mulch of compost or rotted manure around the base of the trunks – keep away from the bark.

# Watering

Newly planted trees must be thoroughly watered when the weather is dry. This need for irrigation is less with established trees, but even here you will have to water if there is a drought. Failure to do so when the tree is in fruit will result in a smaller crop and a reduction in the number of fruit buds next year. Use a hosepipe held close to the ground – apply 50 litres per sq. metre (10 gallons per sq. yard) and repeat every 2 weeks until the dry weather is over.

# Pruning

The purpose of pruning during the first 4 years of the tree's life is **training** – the creation of the basic framework which will ensure satisfactory cropping in later years. With bushes and standards this calls for fairly severe cutting back of the branches in order to produce an open-centred and freely branched tree. In addition you must remove poor quality wood – weak twigs, dead or badly diseased shoots, etc.

After 4 years the purpose of cutting out wood is **maintenance pruning** – the creation of a regular supply of new fruiting wood balanced with the need to retain as much existing fruiting wood as possible. This pruning is generally much less severe than training, but you must continue to remove all dead wood, crossing branches and so on.

All this is winter pruning, designed to encourage shoot growth. However, the main time to prune cordons, espaliers and dwarf pyramids is in summer, and the role of cutting out wood here is to inhibit shoot growth.

### Pruning knife
is useful for cleaning up ragged pruning cuts. Excellent for pruning thin branches, but only if you are experienced in its use

### Long-handled pruner
for stems 1–3 cm (½–1½ in.) across – many gardeners prefer them to a pruning saw for dealing with thicker stems

### Two-bladed secateurs
with proper care will cut cleanly for many years. The cut must be made at the centre of the blades – maximum diameter 1–2 cm (½–¾ in.)

### Pruning saw
is used for branches which are more than 1 cm (½ in.) across – essential for large branches

## Bushes and standards

### Training a 2-year-old tree

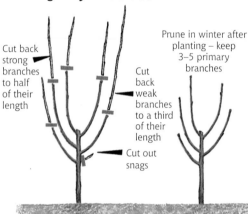

Cut back strong branches to half of their length

Cut back weak branches to a third of their length

Cut out snags

Prune in winter after planting – keep 3–5 primary branches

### Training a 3-year-old tree

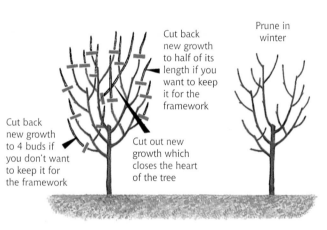

Cut back new growth to 4 buds if you don't want to keep it for the framework

Cut back new growth to half of its length if you want to keep it for the framework

Cut out new growth which closes the heart of the tree

Prune in winter

### Pruning an established tree
The simplest way to look after your mature tree is to follow the Regulation System in winter:

**Spur-bearing variety**
Remove dead and badly diseased wood. Cut back crossing branches and vigorous laterals crowding into the centre.
Then:

**Inside the head** – leave leaders alone. Cut back each lateral which is growing into and beyond the branch leader

**Outside the head** – leave both leaders and laterals alone

Overcropping and undersized fruit may become a problem. If this has happened, thin some of the fruiting spurs and cut out some laterals

**Tip-bearing variety**
Remove dead, diseased and overcrowded wood.
Then:
Cut back some leaders – leave alone all laterals with fruit buds at their tips

# Cordons

## Training

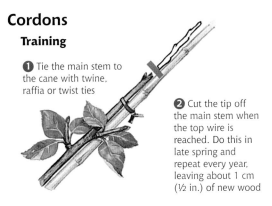

**1** Tie the main stem to the cane with twine, raffia or twist ties

**2** Cut the tip off the main stem when the top wire is reached. Do this in late spring and repeat every year, leaving about 1 cm (½ in.) of new wood

## Pruning an established cordon

Prune in mid-July (southern districts) or early August (other districts) using the Modified Lorette System shown here:

**1** Main stem side-shoot when at least 23 cm (9 in.) long – cut above third leaf beyond basal cluster

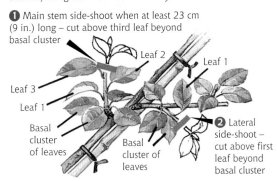

Leaf 2
Leaf 1
Leaf 3
Leaf 1
Basal cluster of leaves
Basal cluster of leaves

**2** Lateral side-shoot – cut above first leaf beyond basal cluster

# Espaliers

## Training

**1** In spring tie 3 canes to the wire supports as shown

**2** In summer train the growth from the terminal bud and 2 side buds along the canes

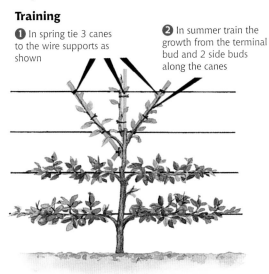

## Pruning an established espalier

Prune in mid-July (southern districts) or early August (other districts). Use the Modified Lorette System described above for cordons. With old espaliers it is often necessary to thin out some overcrowded spurs

**3** In early winter remove the 2 side canes and lower the branches. Tie them carefully along the horizontal wires. Repeat this training process until the final number of tiers is obtained

# Dwarf pyramids

The dwarf pyramid is a compact, high-yielding tree type when grown in open ground, reaching about 2.1 m (7 ft) tall and bearing a distinct Christmas tree shape. It has not become popular with gardeners as it requires careful and regular pruning in summer, but it has proved to be the ideal tree type for pot culture.

# THE PRUNING CUT

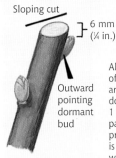

Sloping cut

6 mm (¼ in.)

Outward pointing dormant bud

All pruning cuts must be clean – pare off any ragged parts. Sharp secateurs are essential – press downwards, don't just squeeze. Cuts more than 1 cm (½ in.) across should be painted with a wound sealant to protect them from frost and damp. It is impossible to avoid making some wrong cuts when pruning and as a result snags of dead wood will form above some of the new shoots. Cut off these dead bits as they develop.

# LOPPING

The removal of a large branch from the trunk of a tree. Call in a qualified tree surgeon if the job is larger than your experience.

**2** Saw downwards to sever the main part of the branch

**1** Make a shallow cut on the underside of the branch about 10 cm (4 in.) away from the trunk

Saw off the stub – make the cut almost flush with the trunk

Pare away the rough edges with a pruning knife. Paint the cut surface with a wound sealant

25

# Bird Control

As with all tree fruits, apples have a bird problem. Fruit buds are attacked in winter and the opening blossom is at risk in spring. Cordons in cages are, of course, kept out of harm's way, but virtually all apples are grown in the open. Neither bird scarers nor spray-on repellents are really effective and the range of physical protectants (netting, paper bags around the fruit, cardboard collars around the fruit stalks, etc.) are rarely practical. Apples do not suffer as badly as plums, pears or cherries, and so the usual course of action is to do nothing and trust to luck.

# Picking

The appearance of the first windfalls indicates that the time for picking is near. The colour of the fruit will have started to change and with most varieties the pips will have begun to change from white to brown. However, by far the best guide is to lift the fruit gently in the palm of your hand and give it a slight twist. It is ripe if the apple comes away easily with the stalk attached.

Harvest the fruit at daily intervals rather than trying to gather them all at once. Pick the brightest coloured ones first and handle them gently – place them in a cloth- or straw-lined container. Never tug the apples away from the tree – this can seriously damage the spurs.

Apples gathered for storage should not be over-ripe – if they are, the fruit will rot in store. Most late varieties should be off the tree by late October. Eat or cook bruised or insect-damaged fruit within a day or two – don't try to store them.

# Weed Control

Weed control should begin before planting – pick out and destroy all perennial weed roots when preparing the planting hole. Weeds and grass should be kept away from the stem after planting – a clean ring with a radius of 45–60 cm (18–24 in.) should be maintained. Weeds within this area compete for nutrients, water, etc., and also encourage diseases. Hoeing, hand pulling and mulching are non-chemical answers. The most widely used herbicides are based on natural acids, diquat and glyphosate. Acetic acid, pelargonic acid and diquat burn off the foliage, so keep it off the leaves and avoid drift. It is quick-acting – glyphosate acts more slowly but it does kill or inhibit perennial weeds.

# Storage

Early varieties cannot be stored, but mid-season apples can be kept for a few weeks and the late ones will last for months. Storage life depends on variety and on the conditions – for maximum storage you need a cool, dark and airy store which is not too dry.

There are 2 basic storage techniques. The traditional method is to use a series of trays which are either built or stood on top of each other with space for ventilation between them. You can use wooden orchard boxes, polystyrene containers or inexpensive fibre trays. Wrap each fruit in oiled paper and place folded side down in the tray.

A more recent technique is to store the fruit in clear polythene bags, placing about 2.25 kg (5 lb) of apples in each one. Do not mix varieties. Perforate the bag with a few small holes, bend the top over without excluding all the air and place folded side down in the store. Whichever method you use, remember to inspect the fruit regularly and remove rotten ones immediately.

# Rejuvenating Neglected Trees

❶ The problem often occurs when a keen gardener moves house. In the new garden there are a few old apple trees – crusty-barked, diseased and ugly. A mass of branches form a criss-crossed jungle of wood and leaves, and the fruit is undersized and pest-ridden.

❷ Are the neglected trees worth saving? Even with skill and good luck it will take several seasons to restore productivity, and the work is hard. If the main branches are badly cankered, get rid of them. Only consider saving a tree if the framework is sound and healthy.

❸ The first step in a rejuvenation programme is to consider spacing. Removal of one or more trees will be necessary if overcrowding is a problem, but remember that you may be removing an important pollen provider (see page 14).

❹ The next step is to start to open up and reduce the size of each overgrown tree. It will be necessary to cut out some wood each winter for 3 years – to do all this surgery in a single season would seriously shock the plant.

❺ In the first winter remove all dead, diseased and broken branches. Paint all large pruning cuts with a wound sealant. Next, cut out some of the over-tall branches in the heart of the tree. Also remove some of the branches which are crossing others.

❻ When cutting back a branch do make sure that the piece remaining is large and stout enough to support worthwhile growth in future. Spray with winter wash.

❼ Continue this de-horning (removal or cutting back of stout and over-sized branches) for the next 2 seasons. In addition, you should cut out overcrowded secondary branches to produce an open-centred tree with its main branches spaced about 60 cm (2 ft) apart.

❽ Feed and mulch each season as outlined on page 23. Neglect may mean that yields are still disappointing after the rejuvenation programme. See page 28 for the likely causes of poor yield – take remedial action.

Tree Fruit
Apples

# Causes of Poor Yields

**Impatience** Your newly planted bush cannot be expected to produce fruit for a couple of seasons after planting. Any blossom appearing in the first season should be removed, although letting a single fruit develop for tasting seems to do little harm. Fruiting starts slowly – you will not obtain top yields until the tree is 4–7 years old.

**Poor pruning and careless picking** Strong warnings have appeared on earlier pages. Over-severe pruning of a mature tree will result in abundant new growth in the following summer, and this will be at the expense of fruiting. Cutting off all the tips of laterals is disastrous with tip-bearing varieties. Pulling unripe fruit from the spurs can cause damage and a serious drop in yield next year.

**Overcropping** The tree can only support a limited number of large and well-shaped fruit. A heavy crop of apples should be thinned or the resulting fruit will be small and next year's crop will be very light.

*Wait until the June drop – the time in late June–early July when the tree naturally sheds some of the fruitlets. If the trusses are still crowded, carry out thinning.*
❶ *Remove the king apple (large, central fruit) if it is misshapen*
❷ *Remove small and damaged fruits to leave 2 apples per truss. Use scissors*

**Frost** The effect of a severe frost on open blossom can be disastrous – see page 58. Reduce risk by choosing late-flowering varieties.

**Pests and diseases** Many apple troubles spoil the quality of the fruit, but some lead to lower yields. Typical examples are birds, canker, brown rot and honey fungus. Treat pests and diseases promptly – see pages 52–61.

**Poor location or poor planting** Waterlogging and poor soil will result in poor growth and disappointing yields. There is little you can do with the established tree – tackle the problem before planting and at planting time. Other fruit inhibitors include exposed sites, upland locations and sunless spots.

**Blossom drop and fruit drop** A small amount of these problems will do no harm, but when severe the crop will be seriously reduced. See pages 58 and 61. The absence of a suitable pollination partner is a likely cause.

**Biennial bearing** Some varieties (see the A–Z guide) have an inherent tendency to crop heavily one season and then very lightly 12 months later. However, nearly all other varieties will behave in the same way if not fed, not watered and not thinned.

*If biennial bearing is a problem, try bud rubbing in spring before the expected heavy-cropping season. Rub away about half the fruit buds from the spurs.*

**Over-vigorous growth** Too much nitrogen can result in lush and extensive leaf and stem growth with little or no blossom and fruit formation. Another cause is over-severe pruning. Feeding weekly with a high-potash fertilizer such as tomato food will help, but more drastic action may be necessary. Growing grass around but not up to the trees will help. Cut back some vertical branches in late summer (not winter). Try arching and then tying down some of the branches ('festooning'). If all else fails, try bark ringing.

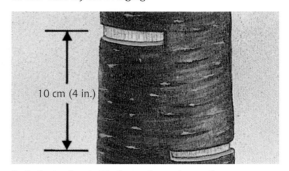

10 cm (4 in.)

*Bark ringing is suitable for apples and pears but not stone fruits. A 6 mm (¼ in.) wide strip of bark is removed in early summer. Cover wound with adhesive tape and paint over with wound sealant. Remove tape in autumn.*

**Starvation** All fruit trees require feeding – a heavy crop is a drain on the soil's reserves. See page 23 for details.

# PEARS

Pears are obviously closely related to apples. They are trained to the same range of tree types and the basic rules of pruning are just the same. The planting technique is identical and many pests and diseases can move from one to the other.

There are many differences, however, and these may be either merely interesting or very important to know. These differences include:

🍐 Pears generally live longer. Fifty years is the normal life span for an apple tree, but a pear should live for a century or more.

🍐 Pears are more temperamental. Dessert types need more sun than apples and they detest cold easterly winds. Young foliage is blackened and torn on exposed sites – you will need some form of windbreak under such conditions.

🍐 Pears have different soil requirements. They will thrive in heavier soil than their more popular rival, but they are less happy in sandy soil, chalky soil and in salt-laden air.

🍐 Pears are more sensitive to frost. Blossom appears 2–4 weeks before apple blooms open, which is a welcome sight but it means that the chance of frost damage is greater.

🍐 Pears usually give lower yields and are slower to come into fruit (4–8 years). Read page 28 if the yields seem to be abnormally low.

🍐 Pears are grown on a more limited range of rootstocks. An apple tree will reach 1.5–7.6 m (5–25 ft) depending on the rootstock. Nearly all pears are grown on quince rootstocks. Quince A is the popular one, producing trees which grow about 3–6 m (10–20 ft) high. Quince C has a dwarfing effect, but this is slight. Trees come into fruit a little earlier and the mature height is about 60 cm (2 ft) less – Quince C also needs a richer soil than Quince A. Pear rootstocks are still used, but only for standards – not for garden cultivation.

🍐 Quince Adams is a newer pear rootstock giving tree size similar to that on Quince C. Trees on this rootstock are high yielding and are planted where the soil is considered too poor for Quince C, or on trees to be grown in containers. Pears are less able to withstand drought but they are generally less prone to pests and diseases. There is one notable exception – birds, such as bullfinches, find the fruit buds of pears more attractive than those of apples.

🍐 Pears have far fewer varieties in the catalogues and at the garden centres. You can expect to find only one third or a quarter of the number of apple varieties in the pear list and suitable choices for most gardeners are very few. Don't pick a cooker – use firm dessert pears instead.

🍐 Pears are more likely to need a pollination partner in your garden. *Conference* is partly self-fertile but it prefers, and most others positively need, a pollination partner – see page 30 for details.

🍐 Pears have a more limited storage life. You can enjoy a wide variety of late apples from store until well into spring, but there is hardly a dessert pear which can be kept until after Christmas.

# Varieties A–Z

You can look through the whole list of varieties if you can provide a warm and sheltered site – if not it would be wise to choose from *Conference*, *Beth* and *Concorde*. Many of the others described here may sound attractive, but they can be unreliable in cooler areas or in a poor season.

Pears are much less popular than apples in gardens, which means that you probably have to provide a suitable pollination partner in your own plot. If space is limited, choose several different cordons. Or you can plant a family tree. The basic type carries the big three – *Conference*, *Doyenné du Comice* and *Williams' Bon Chrétien*.

## POLLINATION GROUPS

| A<br>Early flowering | Louise Bonne de Jersey<br>Packham's Triumph | |
|---|---|---|
| **B**<br>Mid-season flowering | Beth<br>Beurre Hardy<br>Beurre Superfin<br>Conference<br>Durondeau<br>Josephine de Malines | Williams' Bon Chrétien<br><br>**BT**<br>*Triploid – needs two partners*<br>Jargonelle<br>Merton Pride |
| **C**<br>Mid-season/late flowering | Concorde<br>Doyenne du Comice<br>Goldember<br>Gorham<br>Humbug | Invincible<br>Onward<br><br>**CT**<br>*Triploid – needs two partners*<br>Catillac |

Beth

Catillac

### Beth AGM
A new star of the pear world, but not for commercial growers. This excellent English-bred variety is too small and variable in shape for them, but it is a good choice for you. Yields are high and cropping is regular – the white flesh has a melting texture. The upright tree soon comes into fruit. Definitely one for the future.

**Type:** Dessert
**Fruit size:** Medium
**Skin colour:** Olive green. Large patches of russet
**Pollination group:** Partly self-fertile: B
**Picking time:** Late September
**Period of use:** October–November

### Beurre Hardy AGM
Highly recommended for the orchard but not the best choice for a small garden – it is strong-growing, slow to bear fruit and only fairly hardy. *Beurre Hardy* is not always a regular cropper, but the conical fruits are juicy with a distinctive flavour, and the leaves turn bright red in autumn. Pick fruit whilst quite hard – leave it to ripen in store.

**Type:** Dessert
**Fruit size:** Small–medium
**Skin colour:** Pale yellow. Spots of russet
**Pollination group:** B
**Picking time:** Late August
**Period of use:** September

### Beurre Superfin AGM
A variety grown for its better-than-average fruit quality – the white flesh is melting and very sweet. The yields are not as high as some of the catalogues claim and the growth habit is moderate. Do not store for long – the fruit starts to rot once the flesh becomes soft. Shape is long–conical. Scab can be a problem.

**Type:** Dessert
**Fruit size:** Medium–large
**Skin colour:** Greenish-yellow, flushed with red. Large patches of russet
**Pollination group:** B
**Picking time:** Mid-September
**Period of use:** October

### Catillac AGM
A variety with squat, fairly large fruits with a crisp texture. It is not usually eaten fresh, but stores well and the hard flesh cooks to perfection. It forms a large, vigorous tree and is a reliable, heavy cropper with good disease resistance. A very hardy pear, suitable for northern areas, but it is a triploid variety and needs two pollinators.

**Type:** Dessert
**Fruit size:** Medium
**Skin colour:** Golden-yellow. Patches of russet
**Pollination group:** C
**Picking time:** Late September
**Period of use:** October

### Concorde AGM
A newer pear from East Malling Research Station. 'A perfect garden variety' according to one noted nurseryman, but only time will tell. It has the right parents – *Conference* ensures reliability, heavy cropping and an early start to fruiting, and *Doyenne du Comice* provides very good flavour and melting, juicy flesh. Well worth planting in your garden.

**Type:** Culinary
**Fruit size:** Large
**Skin colour:** Pale green
**Pollination group:** C
**Picking time:** October
**Period of use:** February–April

Concorde

**Type:** Dessert

**Fruit size:** Medium

**Skin colour:** Pale green

**Pollination group:** B

**Picking time:** Late October

**Period of use:** November–December

### Conference AGM

This remains the number-one choice because of its reliability under less than perfect conditions. It is self-fertile, but a nearby pollination partner will ensure a better crop. The fruits are long and narrow with juicy but firm flesh. Flavour is satisfactory but not outstanding. Good in the north, but susceptible to scab and wind damage.

---

**Type:** Dessert

**Fruit size:** Medium–large

**Skin colour:** Golden-yellow, flushed brownish-red

**Pollination group:** C

**Picking time:** Mid-October

**Period of use:** November–December

### Doyenne du Comice AGM

The queen of pears if you are looking for flavour – the texture and taste are truly outstanding. A poor choice, however, if you are looking for a reliable tree for less than favourable conditions. It needs warmth, shelter from winds, a nearby pollination partner such as *Beth* or *Concorde* and spraying against scab. An irregular cropper.

---

**Type:** Dessert

**Fruit size:** Medium–large

**Skin colour:** Yellow, flushed red. Patches of russet

**Pollination group:** Partly self-fertile: B

**Picking time:** Late September

**Period of use:** October–November

### Durondeau

The compact growth habit makes this variety a suitable choice for a small garden. It has a good reputation for regularity and heavy cropping, and as a bonus the leaves turn fiery red in autumn. The long fruits store well and the flesh is juicy. Flavour is nothing special. Self-fertile in warm districts. Not really popular despite its charms.

---

**Type:** Dessert

**Fruit size:** Medium

**Skin colour:** Matt brown

**Pollination group:** Self-fertile: C

**Picking time:** Late September–early October

**Period of use:** November–January

### Goldember

('Delsanne') Originally from France, a tree of moderate vigour and upright habit, which starts fruiting at an early age. The fruit is a slightly irregular, spherical shape, with russeted matt-brown skin. The white flesh is smooth and juicy with no grittiness. The distinctive flavour is quite sweet, which is nicely offset by a slight tang of acid.

---

**Type:** Dessert

**Fruit size:** Small–medium

**Skin colour:** Pale yellow. Large patches of russet

**Pollination group:** C

**Picking time:** Early September

**Period of use:** September–October

### Gorham AGM

Quite a popular pear, but there really isn't a great deal going for it. *Gorham* is a regular cropper, but yields are rather light and fruit size is not large. The flavour is good but unusual – 'musky' according to the catalogues. Growth is distinctly upright and spurs form readily. An excellent pollinator. Less prone to scab than its parent *Williams' Bon Chrétien*.

---

**Type:** Culinary/dessert

**Fruit size:** Large

**Skin colour:** Green-, yellow-, pink-striped

**Pollination group:** C

**Picking time:** September

**Period of use:** October–March

### Humbug

('Pysanka') A highly unusual pear with green-, yellow- and pink-striped fruits that have thick skin, firm texture and a sweet, juicy flavour. They can be stored until Easter, making them suitable for both dessert and culinary use. The tree is hardier than most pears and has good disease resistance, so is useful for gardeners in colder areas.

---

**Type:** Culinary/dessert

**Fruit size:** Medium

**Skin colour:** Green to yellow

**Pollination group:** Self-fertile: C

**Picking time:** Late September–early October

**Period of use:** October–February

### Invincible

('Delwinor') Remarkably tough and hardy variety, producing heavy crops of good-quality fruits every year from an early age. The tree will often produce a second flowering after a heavy frost, making it ideal for colder areas. The fruits are large and evenly shaped, changing from emerald green to pale yellow when fully ripe, with sweet, juicy white flesh.

Conference

Doyenne du Comice

Durondeau

Gorham

**Josephine de Malines**

**Louise Bonne de Jersey**

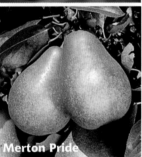

**Merton Pride**

**Williams' Bon Chrétien**

### Jargonelle

The best of the earlies, although the growth habit is straggly. *Jargonelle* is a very old variety which produces good-quality fruit. It is a tip-bearer, so take care when pruning. Good points include high yields and a hardy constitution which makes it suitable for the north – bad points are the very short fruiting season and the lack of fertile pollen.

**Type:** Dessert

**Fruit size:** Medium

**Skin colour:** Greenish-yellow, flushed brownish-red

**Pollination group:** BT

**Period of use:** Early August

**Period of use:** August

### Josephine de Malines AGM

The best of the late varieties – you can enjoy the fruit at the start of the new year. Quality is very good – the pale pink flesh is sweet and perfumed. Yields are high but growth is weak and weeping. It is a tip-bearer with a rather delicate constitution – *Josephine* needs a warm situation in order to ensure a good set of fruit.

**Type:** Dessert

**Fruit size:** Small

**Skin colour:** Greenish-yellow. Russet around stalk area

**Pollination group:** B

**Period of use:** Mid-October

**Period of use:** December–January

### Louise Bonne de Jersey AGM

Rated highly in the flavour league – the flesh is white, melting and sweet. Growth is upright and moderately vigorous – yields are respectable. Blossom appears early, but it is reasonably frost-tolerant. The blossom is attractive and so is the long, colourful fruit. This variety will not pollinate *Williams' Bon Chrétien*.

**Type:** Dessert

**Fruit size:** Small–medium

**Skin colour:** Yellowish-green, flushed and spotted with red

**Pollination group:** A

**Period of use:** Late September

**Period of use:** October

### Merton Pride

A high-quality pear – the creamy flesh is juicy and the flavour is excellent. A good cropper, but *Merton Pride* sometimes becomes a biennial bearer – see page 27. Check that the plant has been 'double worked' to ensure compatibility with the quince rootstock. The fruits are conical and this variety appears in many recommended lists.

**Type:** Dessert

**Fruit size:** Medium–large

**Skin colour:** Pale green. Patches of russet

**Pollination group:** BT

**Period of use:** Early September

**Period of use:** September

### Onward AGM

A newer pear which is highly regarded. *Doyenne du Comice* is one of the parents, but the two varieties will not pollinate each other. Fruit quality is excellent. The problem with *Onward* is the short period of use – storage qualities are poor which means that the crop is finished by early October. Still, an excellent modern variety.

**Type:** Dessert

**Fruit size:** Medium–large

**Skin colour:** Yellowish-green, flushed with pink. Patches of russet

**Pollination group:** C

**Period of use:** Mid-September

**Period of use:** Late September

### Packham's Triumph

A *Williams'*-type pear which crops later in the season. Both cropping and keeping qualities are good – the conical, irregular-shaped fruit is very juicy and sweet. Tree growth is compact and upright – a reasonable choice but the early-flowering habit makes it susceptible to frost. A variety recommended for milder districts.

**Type:** Dessert

**Fruit size:** Medium

**Skin colour:** Yellow, flushed with orange

**Pollination group:** A

**Period of use:** Mid-October

**Period of use:** November–December

### Williams' Bon Chrétien AGM

Popular as a shop-bought fruit and as *Bartlett* pears in tins. As a garden tree this old English variety is a regular cropper and hardy enough for northern districts. The quality of the oval, smooth-skinned fruit is very good but storage qualities are poor and so is disease resistance. Will not pollinate *Louise Bonne de Jersey*.

**Type:** Dessert

**Fruit size:** Medium–large

**Skin colour:** Pale yellow, striped and spotted with pale red

**Pollination group:** B

**Period of use:** September

**Period of use:** September

# Planting and Pruning

The site for pears should be chosen with a little more care than for apples. Some shelter from cold winds is essential. In northern districts choose a spot close to a wall – the microclimate will be warmer and less windy. Follow the basic rules for planting set out on pages 9–10. Plant to the old soil mark – the union with the rootstock should be about 10 cm (4 in.) above ground level.

Pears can be grown as bushes, dwarf pyramids, cordons, espaliers and fans in the same way as apples. For pruning details see pages 24–25. The usual practice is to buy bushes as 2-year-old trees and supported types such as cordons as 3-year-old ones. Very few nurseries name the rootstock on the label – it is most likely to be Quince A. Do not plant standard or half-standard pear trees in an average-sized garden – they are far too vigorous and far too much trouble.

| Tree type | Space between trees | Space between rows | Yield per mature tree |
|---|---|---|---|
| Standard | 9 m (30 ft) | 9 m (30 ft) | 45–112 kg (100–250 lb) |
| Half-standard | 6 m (20 ft) | 6 m (20 ft) | 23–56 kg (50–125 lb) |
| Fan | 4.5 m (15 ft) | – | 5.5–10.5 kg (12–24 lb) |
| Bush | 4.5 m (15 ft) | 4.5 m (15 ft) | 23–45 kg (50–100 lb) |
| Espalier | 4.2 m (14 ft) | 1.8 m (6 ft) | 7–11 kg (15–25 lb) |
| Dwarf bush | 2.4 m (8 ft) | 4.5 m (15 ft) | 9–18 kg (20–40 lb) |
| Spindlebush | 2.1 m (7 ft) | 4.2 m (14 ft) | 18–36 kg (40–80 lb) |
| Dwarf pyramid | 4.5 m (15 ft) | 2.4 m (8 ft) | 4.2–5.5 kg (8–12 lb) |
| Cordon | 75 cm (2½ ft) | 1.8 m (6 ft) | 2.1–2.7 kg (4–6 lb) |

*Use a cane to make sure the graft union is above soil level.*

# Windbreaks

Pears flower earlier than apples, and at this time few pollinating insects are present. Turbulent air around the trees will deter them, so you should consider a windbreak if the site is exposed. A solid wall or fence over which the wind can blow will actually increase air turbulence – use a plastic windbreak through which the air can pass. A nearby hedge will provide a useful windbreak – it will gently reduce the wind speed for a distance of 15–30 times the height of the hedge.

## Feeding and Mulching See page 23.

## Watering See page 23.

## Bird Control See page 26.

## Weed Control See page 26.

# Maypoling

A heavy crop can cause one or more branches to break. Support heavily laden branches by propping them up with a forked pole or by maypoling as shown below.

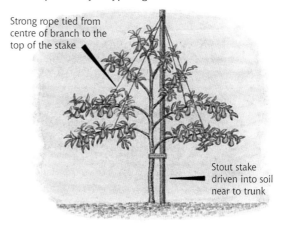

Strong rope tied from centre of branch to the top of the stake

Stout stake driven into soil near to trunk

# Picking

A pear is ripe if it readily parts from the tree when lifted gently in the palm of the hand and given a slight twist. Skill is required to judge the correct time of picking. With early-ripening varieties the fruits should be removed when they are full-sized but before they have reached the fully ripe stage described above. Cut the stalks and leave the fruits for a few days for the full flavour to develop. Eat as they ripen – do not attempt to store. Later varieties are ripened in store. Pick when they come away from the tree quite easily.

# Storage

You can follow the general rules described for apple storage on page 26, but the experts prefer a slightly different technique. Pears are best stored in slatted trays and they should not be wrapped. The stalks should be intact and the need for careful handling and the rejection of bruised or damaged fruit is even greater than with apples. The atmosphere around the stored fruit can be drier and warmer than in the apple store – 7°C (45°F) is quite acceptable. Ripeness is indicated by a softening near the stalk. Bring the ripe fruit into the kitchen for a few days to complete the ripening process.

# PLUMS

Plums are the most popular of the stone fruits and they are also the easiest to grow. However, plums are not for everyone. They flower very early in the season and that means cropping can be very disappointing and irregular in low-lying or exposed sites. Choose the highest spot in the garden and a position which gets lots of sun if you plan to grow a dessert plum or gage.

A standard or half-standard has no place in the ordinary garden. Even a bush or pyramid on the popular St Julien A rootstock can reach 6 m (20 ft) or more when mature. This would be a problem in a small garden. One way of tackling the problem would be to choose a specimen grafted on to Pixy (see below). Perhaps the best way of saving space is to grow the plum as a fan against a south-facing wall. Note that plums are not grown as espaliers or cordons. An added advantage of the fan growth form is the ease with which netting can be used to protect the buds and fruit from birds.

Cropping should start when the tree is about 5 years old – earlier if Pixy rootstock is used. The mature tree is easily cared for, but the dreaded silver leaf disease is a constant threat. Choose your plant with care – some fruits are sweet and others are sour. There are several distinct types as described below.

| Mature bush 5–8 m (15–25 ft) high | **Dessert plum** Sweet – eaten fresh. Fleshy fruit – trees smaller and less hardy than culinary varieties. Most popular variety: *Victoria* | **Culinary plum** Rather tart – used for cooking. Fruit less fleshy than dessert varieties – trees more tolerant of poor conditions. Most popular variety: *Czar* | **Gage** Smaller, rounder and sweeter than dessert plums. Yields are not high. Rather tender – grow as a fan against a south wall |
| --- | --- | --- | --- |
| Mature bush 3–5 m (10–15) ft high | **Bullace** Sharp flavour – used for cooking. Fruit left until late autumn before picking. Choose a damson variety instead | **Damson** Spicy tart flavour – used for cooking, jam- and wine-making. A hardy tree which succeeds where a plum would fail | **Mirabelle** Similar in shape and size to bullace, but sweeter and golden-yellow. Is now becoming more popular and is offered by a small number of specialist fruit nurseries |
| Mature bush 6–10 m (20–30 ft) high | **Myrobalan (cherry plum)** A decorative plum used as a specimen tree and for hedging – attractive blossom in March. Cherry-like fruit in July which can be used for cooking or jam-making | | |

# Rootstocks

The most widely used rootstock is St Julien A – this is the one to choose if conditions are less than ideal. A newer dwarfing rootstock called Pixy is quite popular, as it produces bushes with a mature height of only 3–3.6 m (10–12 ft) and the trees begin cropping very early in life (usually about 3 years). You do need good soil and good growing conditions for Pixy-rooted plums and it is also recommended for plants intended for growing in containers, although some nurserymen do not recommend this rootstock for northern districts. More recent introductions are VVA1, which is dwarfing and will reach a height of 2.1–2.7 m (7–9 ft) when mature. Trees usually start bearing fruit within 3 years.

Other stocks are available. Myrobalan B is far too vigorous for ordinary gardens. Brompton comes to fruit quickly and resists suckering, but it is much more vigorous than St Julien A. Mussel produces a more compact tree than either Myrobalan B or Brompton.

# Varieties A–Z

Plums are one of the earliest fruit trees to open their flowers in the spring, which means that pollination is sometimes disappointing. Hand pollination of a well-grown plum is worthwhile – see page 43.

Many plums are self-fertile, including the most popular dessert variety (*Victoria*) and the most widely grown cooker (*Czar*). Others need a pollination partner chosen from the same or an adjacent group – see the table on the right. Buy 2- or 3-year-old partly trained trees and if the site is unfavourable choose varieties in the late-flowering group. *Victoria* remains a good overall choice – it is reliable and flavour is acceptable, but it is susceptible to silver leaf disease.

## POLLINATION GROUPS

| | | |
|---|---|---|
| **A** Early flowering | Farleigh Damson Warwickshire Drooper | |
| **B** Mid-season flowering | Czar Denniston's Superb Herman Merryweather Opal | Pershore Yellow Purple Pershore Rivers' Early Prolific Sanctus Hubertus Victoria |
| **C** Late flowering | Bullace Cambridge Gage Early Transparent Gage | Kirke's Blue Marjorie's Seedling Oullin's Golden Gage Shropshire Prune Damson |

**Tree Fruit** / **Plums**

---

**Type:** Bullace
**Use:** Culinary
**Fruit size:** Very small
**Skin colour:** Blue-black
**Pollination group:** Self-fertile: C
**Period of use:** November

### Bullace
Search all the catalogues and you will find a supplier, but it really isn't worth growing. The yellow variety (*Shepherd's Bullace*) is even more difficult to find. The sharp-flavoured fruits have to be left on the thorny branches until late autumn in order to allow early frosts to mellow the acidity. The wild form grows in hedgerows.

**Bullace**

---

**Type:** Gage
**Use:** Dessert/culinary
**Fruit size:** Small
**Skin colour:** Yellowish-green, flushed with red
**Pollination group:** Partly self-fertile: C
**Period of use:** Early September

### Cambridge Gage
Very similar to the old *Greengage* but more reliable and a heavier cropper. The flesh is yellowish-green and juicy – flavour is very good. Growth is vigorous, but this variety cannot be recommended for cold areas. Unlike the traditional *Greengage* it does set some fruit without a partner. You will find *Cambridge Gage* in nearly all the catalogues.

**Cambridge Gage**

---

**Type:** Plum
**Use:** Culinary
**Fruit size:** Medium
**Skin colour:** Dark purple
**Pollination group:** Self-fertile: B
**Period of use:** Early August

### Czar
The cooking variety most people choose – the oval fruits have yellowish-green flesh which has a good acidic flavour. *Czar* can be used as a dessert plum when fully ripe. It is one of the most reliable varieties in less-than-perfect areas as its frost resistance is high, but *Czar* has low disease resistance. Growth is upright.

---

**Type:** Gage
**Use:** Dessert
**Fruit size:** Large
**Skin colour:** Greenish-yellow, streaked with dark green
**Pollination group:** Self-fertile: B
**Period of use:** Late August

### Denniston's Superb
Listed with the gages but it is really a plum. *Denniston's Superb* has a good reputation for hardiness and reliability – growth is upright and vigorous. The transparent flesh is sweet and the skin often has a red flush. Yields are high and this variety is widely recommended for Midland and northern regions.

---

**Type:** Gage
**Use:** Dessert
**Fruit size:** Small
**Skin colour:** Yellow, spotted with red
**Pollination group:** Self-fertile: C
**Period of use:** August

### Early Transparent Gage
Choose this one for flavour – the golden, melting flesh is very sweet. Don't expect bumper crops – growth is neat and upright, and the round fruits are small. Still, it is a reliable variety and a regular cropper – no years of plenty and then famine with this one. *Early Transparent Gage* is available from specialist fruit nurseries.

**Czar**

35

**Herman**

**Marjorie's Seedling**

**Merryweather**

**Opal**

**Oullin's Golden Gage**

### Farleigh Damson
The small oval fruits are dark, metallic blue, and are blotched with black with a light blue bloom. They have a strong, sharp damson flavour. The tree has a compact habit with a dense growth of horny branches, and is a prolific and regular cropper, producing dense clusters of fruit that can cause the branches to droop downwards.

**Type:** Damson
**Use:** Culinary
**Fruit size:** Small
**Skin colour:** Blue-black
**Pollination group:** Partially self-fertile: A
**Period of use:** Mid-September

### Herman
A very early, medium-sized, blue-black plum with outstanding quality and a good flavour. The yellow flesh parts from the stone very easily. Bred in Sweden, this plum grows well in northern areas of the UK. Although it is usually ready for picking in late July, it does not keep well. The tree has an upright–rounded habit and will tolerate most soils, except very chalky or badly drained.

**Type:** Plum
**Use:** Culinary/dessert
**Fruit size:** Medium
**Skin colour:** Blue-black
**Pollination group:** Self-fertile: B
**Period of use:** Late July

### Kirke's Blue
All the suppliers sing the praises of the eating qualities of this dark-coloured plum. The greenish-yellow flesh is firm but juicy with an outstandingly sweet flavour. The tree is easy to train as growth is compact, but there are drawbacks. Yields are often disappointing and it needs a sheltered spot in the garden if it is to thrive.

**Type:** Plum
**Use:** Dessert
**Fruit size:** Large
**Skin colour:** Dark purple
**Pollination group:** C
**Period of use:** Early September

### Marjorie's Seedling
*Marjorie's Seedling* is an excellent choice if you want a late plum which can be used for cooking and eating fresh. The oval fruit has yellow flesh and is moderately sweet when fully ripe. Growth is both vigorous and upright, and the blossom season is late enough to miss early frosts. Picking starts at the end of September but can continue until mid-October.

**Type:** Plum
**Use:** Dessert/culinary
**Fruit size:** Large
**Skin colour:** Purple
**Pollination group:** Self-fertile: C
**Period of use:** Late September

### Merryweather
This damson has the characteristic features you would expect from this group – blue-black skin, yellow flesh which is juicy and acidic, and a hardy constitution. *Merryweather* is an excellent all-round performer and you will find it in most catalogues, but the other popular damson (*Shropshire Prune Damson*) is a better choice where space is limited.

**Type:** Damson
**Use:** Culinary
**Fruit size:** Large for a damson
**Skin colour:** Blue-black
**Pollination group:** Self-fertile: B
**Period of use:** Late September

### Opal
A more recent variety, which is gradually replacing *Early Laxton* for the most popular early plum crown. *Opal*'s flavour is better – there is a distinct gage-like texture and taste. A feature noted in the catalogues is the ease with which the yellow flesh parts from the stone. It is easy to manage but the buds are extremely attractive to birds during winter.

**Type:** Plum
**Use:** Dessert
**Fruit size:** Medium
**Skin colour:** Reddish-purple
**Pollination group:** Self-fertile: B
**Period of use:** Late July

### Oullin's Golden Gage
The best choice of all, according to some experts, but it is not a true gage – it is a gage-like plum. The fruit is large and round and the yellow transparent flesh is moderately but not outstandingly sweet. The fruit is excellent for bottling and freezing as well as for eating fresh. Growth is vigorous, upright and reasonably healthy.

**Type:** Gage
**Use:** Dessert/culinary
**Fruit size:** Large
**Skin colour:** Golden-yellow, spotted with green or brown
**Pollination group:** Self-fertile: C
**Period of use:** Mid-August

**Type:** Plum
**Use:** Culinary
**Fruit size:** Medium
**Skin colour:** Golden-yellow
**Pollination group:** Self-fertile: B
**Period of use:** Mid-August

### Pershore Yellow

Here is the one for yellow plum jam and golden pie filling. An excellent cooker, but the flavour is rather insipid for eating fresh. The tree is one of the most reliable you can choose. Advantages include good frost resistance, regular cropping and heavy yields. It is self-fertile, so no pollination partner has to be provided.

**Type:** Plum
**Use:** Culinary
**Fruit size:** Medium
**Skin colour:** Bluish-purple
**Pollination group:** Self-fertile: B
**Period of use:** Mid-August

### Purple Pershore

The purple equivalent of *Pershore Yellow*. The same excellent cooking qualities are there, but this time you get purple stewed fruit, jam or tart filling. Some experts feel that the flavour is superior to the yellow variety. The same desirable growth properties are found – good disease resistance, prolific cropping and reliability.

**Purple Pershore**

**Type:** Plum
**Use:** Culinary
**Fruit size:** Small
**Skin colour:** Bluish-purple
**Pollination group:** B
**Period of use:** Early August

### Rivers' Early Prolific

Everyone agrees that this rather small and spreading tree lives up to its name – it is both early and prolific. The round or oval fruits are quite small but they are golden-fleshed and juicy with a moderately good flavour. It can be used as a dessert variety when the fruit is fully ripe. The best or worst of the earlies – the experts can't agree.

**Type:** Plum
**Use:** Dessert/culinary
**Fruit size:** Medium
**Skin colour:** Bluish-purple
**Pollination group:** Partly self-fertile: B
**Period of use:** Early August

### Sanctus Hubertus

A newer variety from Belgium which is gradually taking over from *Rivers' Early Prolific*. The fruit is larger and there is a heavy grey bloom. The flowers are partly self-fertile. The flavour is good enough to make it a dual-purpose plum and the tree is easy to manage. Unfortunately there are very few suppliers, but this should change.

**Type:** Damson
**Use:** Culinary
**Fruit size:** Small
**Skin colour:** Dark blue
**Pollination group:** Self-fertile: C
**Period of use:** Early September

### Shropshire Prune Damson

Small, oval fruit of blue-black colour with a dense bloom. A classic damson considered to be one of the best, with a strong, rich, astringent damson flavour. It forms a compact tree with a dense habit and thicket of twiggy branches. Fair and regular cropper rather than producing very heavy crops. Grows well in northern areas and in high rainfall.

**Victoria**

**Type:** Plum
**Use:** Dessert/culinary
**Fruit size:** Large
**Skin colour:** Pale red
**Pollination group:** Self-fertile: B
**Period of use:** Late August

### Victoria

Still the most popular plum with both amateur and commercial growers. Flesh is yellowish-green. 'If you can grow only one, this is it,' say many catalogues. However, do think twice. It is a reliable and heavy cropper, but thinning is often essential to avoid biennial bearing. Its disease resistance is poor and flavour is only average.

**Type:** Plum
**Use:** Dessert
**Fruit size:** Large
**Skin colour:** Yellow
**Pollination group:** Self-fertile: A
**Period of use:** Late September

### Warwickshire Drooper

A vigorous tree with an attractive weeping habit. A heavy, regular cropper. Very large, oval–oblong, yellow fruits with a thick skin speckled with red spots, brownish russet patches and covered with a thin grey bloom in mid–late September. The yellow flesh is tender and quite sweet. Fruit hangs well on the tree. Superb for all culinary purposes and can also be eaten for dessert and freezes well.

**Warwickshire Drooper**

# Planting

Plums need a moisture-retentive but free-draining soil. Light soils which easily dry out must be enriched with humus. November is the best time for planting. Staking will be required for 5–6 years.

*Drive a stout stake in firmly.*

| Tree Type | Space between trees | Space between rows | Yield per mature tree |
|---|---|---|---|
| Standard | 7.6 m (25 ft) | 7.6 m (25 ft) | 22.5–45 kg (50–100 lb) |
| Half-standard | 6 m (20 ft) | 6 m (20 ft) | 13.5–27 kg (30–60 lb) |
| Fan | 4.5 m (15 ft) | – | 5.5–11 kg (12–24 lb) |
| Bush | 4.2 m (14 ft) | 4.2 m (14 ft) | 14–22.5 kg (30–50 lb) |
| Pyramid | 3.6 m (12 ft) | 3.6 m (12 ft) | 14–22.5 kg (30–50 lb) |

# Pruning

There are two basic points to remember – do not prune in winter and paint all cuts with wound sealant. These measures are necessary to reduce the risk of silver leaf infection.

## Bushes

### Training a 2- or 3-year-old tree

Prune in March. The simplest procedure is to follow the plan for apples – see page 24. The goal is to have 3–5 strong branches at the 2-year-old stage – these branches should be as near to horizontal as possible. At the 3-year stage there should be about 8 strong primary and secondary branches which are well spaced.

### Pruning an established tree

Prune in June–late July. Keep pruning to a minimum. The sole purpose of cutting out wood at this stage is to keep the tree healthy and to reduce overcrowding. Remove dead, dying, broken and diseased branches. Cut back overcrowded branches. Suckers growing around the base should be removed at this stage. Pull them up – do not cut off with secateurs.

## Fans

Create the framework of the tree by following the instructions for peaches – see page 46.

In spring rub away buds growing towards or away from the wall

In July pinch the side shoots back to 6 leaves

After harvest cut back the side-shoots to half their length

# Seasonal Care

Bird damage can be severe. Both buds in spring and fruit in summer are at risk. Cover the tree with netting if practical.

Water regularly in dry weather – apply up to 25 litres per sq. metre (5 gallons per sq. yard) every 10 days until the dry weather breaks. Avoid heavy watering at irregular intervals, as fruit splitting can result.

Hoe to keep down weeds around the base of the tree, but avoid hoeing deeply at all costs. Surface root damage will result in sucker formation.

Fruit overcrowding results in small, tasteless fruit and biennial bearing. Thin with scissors in early June and again at the end of the month if the crop is heavy – leave fruit 8–10 cm (3–4 in.) apart. Support overladen branches by post support or by maypoling – see page 33.

# Feeding and Mulching

See page 23. Plums require more nitrogen than apples – give a late-spring application of nitrogen-rich liquid fertilizer such as Growmore or pelleted chicken manure.

# Picking and Storing

A plum is ripe when it parts easily from the tree. With plums the stalk usually remains on the branch – with gages and damsons the stalk comes away with the fruit. Pick dessert varieties when thoroughly ripe – go over the tree several times. Culinary varieties should be harvested while the fruit is still slightly unripe – use within 1–2 weeks.

Dessert plums will keep for only a few days. They can be kept for 2–3 weeks in a cool place if picked when still unripe and with the stalks attached. Remove stones from plums before freezing.

# CHERRIES

Until quite recently there was no point in trying to grow a sweet cherry in an average-sized garden. The spring blossom was attractive, but the tree would grow steadily until it became a monster with a spread and height of about 12 m (40 ft). This meant that proper maintenance was extremely difficult, and the need for a pollination partner meant that two trees had to be grown. At least the chore of picking was unnecessary – without protection the fruit merely served as living bird food.

Things began to change in the late 1970s when the self-fertile *Stella* was introduced. At about the same time Colt rootstock enabled trees to be maintained at 4.5–6 m (15–20 ft). Grown as a pyramid it is possible to restrict growth to about 3 m (10 ft) and it is even practical to grow *Stella* grafted on to a Colt in a pot on the patio. For care of pot-grown fruit trees, see Chapter 4. Prune in April instead of winter – see page 42.

The speed of change has begun to quicken. *Compact Stella* has been introduced – a less vigorous sport of *Stella*. New self-fertile varieties have come over from Canada – you will find *Cherokee* and *Sunburst* in one or more specialist catalogues. Even more interesting is the dwarfing rootstock Inmil (GM9). The promise is for trees that grow to a maximum height of 1.8–2.1 m (6–7 ft) – and that means netting the branches and reaching the fruit will no longer be a problem. Cherries on this rootstock are beginning to appear in the catalogues, but some growers are awaiting the outcome of further trials before introducing Inmil-grafted cherries.

Until the sweet cherry revolution described above, it was traditional to choose an acid cherry variety for garden cultivation. Growth is much more compact and the most popular variety (*Morello*) is self-fertile. Trees begin to bear fruit after only 3 or 4 years and the acid cherries grow quite happily in some shade. This means that *Morello* can be trained as a fan against a north-facing wall.

So cherry-growing is a practical proposition nowadays, and you don't have to restrict yourself to an acid variety. All you need for a sweet cherry fan is a 2.4 m (8 ft) high wall in full sun and the application of netting when birds are active.

### Sweet (dessert) cherry
Sweet – eaten fresh. Flesh is either soft and very juicy or firm and moist. Skin colour ranges from yellow to near black. Traditionally very vigorous and self-incompatible, but semi-vigorous and self-fertile trees are now available

### Duke cherry
A cross between the sweet and acid cherry – the flavour is intermediate between the two. Flesh is juicy. *May Duke* is the only variety you are likely to find.

### Acid (sour or culinary) cherry
Sour – used for cooking, bottling, jam, wine-making, etc. Flesh may be white, pink or red. *Morello* is the usual choice, but other self-fertile acid varieties such as *Nabella* are available.

## Rootstocks

Until the mid-1970s we had to make do with the vigorous rootstock Malling F12/1. Trees grown on this stock require a lot of space and you have to wait a number of years before fruiting begins. The introduction of the dwarfing rootstock Colt has resulted in trees which grow to only about half the height of an F12/1-grafted cherry. Not a dwarf tree, but still practical as a fan for the average garden. More recently, Gisela 5, an even more dwarfing rootstock (only about 40% of height of Colt), has become available and is ideal for gardens and patio pots.

# Varieties A–Z

Choose a self-fertile variety. Pick the well-established *Stella* if you want a sweet variety, or you can choose one of the new ones – *Compact Stella* or *Sunburst*. *Morello* is the acid cherry chosen by nearly everyone, although you could try the recently introduced *Nabella*.

Other varieties require a pollination partner, and the situation with cherries is complex. Take care to pick a suitable partner – it will have to flower at approximately the same time and it will also have to be compatible, so with cherries a specific partner is usually named.

## POLLINATION PARTNERS

*Stella, Noir de Guben* or *Merchant* can be used to pollinate nearly all the varieties listed below. Specific pollinators, where one is needed, are shown here:

| Variety | Pollen Provider |
| --- | --- |
| Early Rivers | Merton Glory, Noir de Guben |
| Kordia | Lapins, Penny |
| Lapins | Self-fertile |
| May Duke | Partly self-fertile |
| Merchant | Early Rivers, Merton Glory |
| Merton Glory | Early Rivers, Morello |
| Morello | Self-fertile |
| Napoleon Bigarreau | Morello, Stella |
| Penny | Lapins, Summer Sun |
| Stella | Self-fertile |
| Summer Sun | Self-fertile |
| Sunburst | Self-fertile |
| Sweetheart | Self-fertile |
| Van | Merton Glory, Stella |

Kordia

Merchant

Merton Glory

### Early Rivers

Introduced over 100 years ago and still popular. A black cherry noted for its earliness and its fruit size. Both skin and flesh are very dark and flavour is excellent. Cropping is heavy and regular, but its vigour and spreading habit can be a problem.

**Type:** Sweet

**Fruit size:** Very large

**Skin colour:** Near black

**Picking time:** Mid-June

### Kordia AGM

A newer variety, originating from Czechoslovakia, which is being recommended as a good possibility for more northerly areas. A heavy-cropping, true black dessert cherry with large, deep red, heart-shaped fruits, with a deep purple flesh that is sweet and juicy. The fruits are resistant to splitting and the variety has good disease resistance.

**Type:** Sweet

**Fruit size:** Large

**Skin colour:** Dark red

**Picking time:** Mid-August

### Lapins AGM

Also known as *Cherokee*, this forms an upright and strong-growing tree. If there is a very heavy crop the tree will shed some of its fruit when green to thin the fruits naturally. The black fruit has dark red flesh and an exquisite flavour, with good resistance to splitting.

**Type:** Sweet

**Fruit size:** Large

**Skin colour:** Black

**Picking time:** Mid-July

### May Duke

Generally agreed to be the best of the duke cherries. It is compact with only moderate vigour like an acid variety, but it should be pruned in the same way as a sweet cherry. The flavour is tart but not unpleasant. Cropping is only moderate.

**Type:** Duke

**Fruit size:** Large

**Skin colour:** Dark red

**Picking time:** Mid-July

### Merchant AGM

A newer variety with a number of virtues. The fruit is large and the yields are heavy, and earlier than many varieties, but prone to splitting. A tree of moderate vigour with some resistance to bacterial canker. It is not self-fertile, but it will pollinate all the other varieties listed here.

**Type:** Sweet

**Fruit size:** Very large

**Skin colour:** Dark red or near black

**Picking time:** Early July

### Merton Glory

This is the one to choose if you want an outstandingly large yellow cherry early in the season. The white flesh is firm and the fruit is heart-shaped rather than round. Growth is distinctly upright and trees are quite widely available.

**Type:** Sweet

**Fruit size:** Very large

**Skin colour:** Yellow, flushed red

**Picking time:** Late June

### Morello

**Type:** Acid

**Fruit size:** Large

**Skin colour:** Near black

**Picking time:** August

By far the most popular acid cherry – until recently the only self-fertile one. It can be picked at the very sour, dark red stage for stewing or it can be left under netting until near black and bitter-sweet. The flesh is dark red – growth is compact and spreading.

### Napoleon Bigarreau

**Type:** Sweet

**Fruit size:** Medium–large

**Skin colour:** Pale yellow, flushed with dark red

**Picking time:** Early August

One of the great Bigarreau or fleshy varieties, introduced into Britain over 150 years ago. Renowned for generations for its firm, very sweet flesh. It is a heavy cropper and you will have no trouble in finding a supplier. Sometimes sold as *Bigarreau Napoleon*.

### Penny AGM

**Type:** Sweet

**Fruit size:** Large

**Skin colour:** Black

**Picking time:** Late August

This recently introduced, late-season cherry, bred in the UK, is a very regular cropper. The dark red fruits gradually turn black as they ripen and are exceptionally large, firm and well flavoured. Crops heavily and regularly from an early age.

### Stella

**Type:** Sweet

**Fruit size:** Large

**Skin colour:** Dark red

**Picking time:** Late July

The first self-fertile sweet cherry – the number-one choice for garden use. Growth is vigorous – upright at first and later spreading. The flavour is good and the flesh is juicy. A good pollinator for other cherries. Fruit may be dark red or near black.

### Summer Sun AGM

**Type:** Sweet

**Fruit size:** Large

**Skin colour:** Dark red

**Picking time:** Mid–late July

This particularly hardy cherry can be relied on to produce good crops in more exposed, colder areas of the UK due to its frost-hardiness. Its compact bushy form makes it an ideal candidate for the smaller garden or for where space is limited. A reliable and heavy cropper producing large fruits with a delicious sweet flavour.

### Sunburst

**Type:** Sweet

**Fruit size:** Large

**Skin colour:** Near black

**Picking time:** Early July

A new one from British Columbia – the first self-fertile and black sweet cherry. Similar to *Stella* in general growth habit, but the fruit is much darker. The UK supplier claims 'a gorgeous flavour'. Try it and see – you won't need a pollination partner.

### Sweetheart AGM

**Type:** Sweet

**Fruit size:** Medium

**Skin colour:** Deep red

**Picking time:** Mid-August

A very productive, moderately vigorous tree with an upright, spreading habit, but with some susceptibility to mildew. Good-quality, deep red, heart-shaped fruits with a good, sweet flavour, ripening over an extended period. A self-fertile and regular, heavy-cropping variety.

### Van

**Type:** Sweet

**Fruit size:** Large

**Skin colour:** Dark red

**Picking time:** Late July

This variety is chosen for its reliability. Cropping is both regular and heavy, but growth is very vigorous. Not a good choice where space is limited. The dark red flesh is firm – the flavour is sweet and some consider it to be superb.

Morello

Napoleon Bigarreau

Stella

Sweetheart

# Planting

Sweet cherries need a sheltered site in full sun – acid varieties are less fussy. Both need deep soil which is moisture-retentive but free-draining. Don't try to grow cherries in shallow or sandy soil; grow them in containers instead. November is the best time to plant.

*Diagonal support stakes are used for container-grown trees.*

| Tree type | Cherry type | Rootstock | Space between trees | Space between rows | Yield per mature tree |
|---|---|---|---|---|---|
| Standard | Sweet | F12/1 | 10.6 m (35 ft) | 10.6 m (35 ft) | 45 kg (100 lb) |
| Half-standard | Sweet | F12/1 | 7.6 m (25 ft) | 7.6 m (25 ft) | 32 kg (70 lb) |
| Bush | Sweet | F12/1 | 7.6 m (25 ft) | 7.6 m (25 ft) | 23 kg (50 lb) |
| Fan | Sweet | F12/1 | 6 m (20 ft) | – | 11 kg (25 lb) |
| Bush | Sweet | Colt | 4.5 m (15 ft) | 4.5 m (15 ft) | 16 kg (35 lb) |
| Fan | Sweet | Colt | 4.5 m (15 ft) | – | 9 kg (20 lb) |
| Bush | Sweet | Gisela 5 | 3.6 m (12 ft) | 3.6 m (12 ft) | 18 kg (40 lb) |
| Fan | Sweet | Gisela 5 | 3 m (10 ft) | – | 14 kg (30 lb) |
| Bush | Acid | F12/1 | 6 m (20 ft) | 6 m (20 ft) | 18 kg (40 lb) |
| Fan | Acid | F12/1 | 6 m (20 ft) | – | 9 kg (20 lb) |
| Bush | Acid | Colt | 4.5 m (15 ft) | 4.5 m (15 ft) | 14 kg (30 lb) |
| Fan | Acid | Colt | 4.5 m (15 ft) | – | 9 kg (20 lb) |
| Bush | Acid | Gisela 5 | 3.6 m (12 ft) | 3.6 m (12 ft) | 20 kg (45 lb) |
| Fan | Acid | Gisela 5 | 3 m (10 ft) | – | 16 kg (35 lb) |
| Pyramid | Sweet | Colt | 3.6 m (12 ft) | 3.6 m (12 ft) | 11 kg (25 lb) |
| Pyramid | Sweet | Gisela 5 | 2.7 m (9 ft) | 3 m (10 ft) | 14 kg (30 lb) |

# Pruning

Do not prune in winter and paint all cuts with wound sealant. These measures are necessary to reduce the risk of silver leaf and bacterial canker infection.

### Sweet cherry: bushes

Follow the rules laid down for plum bushes (page 38) for both the training of young trees and the pruning of established trees.

### Acid cherry: bushes

Follow the training procedure given for plum bushes – see page 38. With established trees remove dead, diseased and damaged branches in spring but leave healthy old wood. Fruit will only develop on young shoots, so each year after harvest remove some of the poor-yielding branches in order to stimulate new growth.

### Sweet cherry: fans

Follow the training procedure given for peach fans – see page 46. With established fans follow the diagrams on page 38. In April rub away unnecessary buds and in July pinch back the tips of new shoots. In September shorten these shoots again, leaving 3 or 4 buds.

### Acid cherry: fans

Follow the training procedure given for peach fans – see page 46. With established fans remove inward- or outward-facing buds in April. Fruit will only develop on young shoots so each year after harvest cut back each fruited stem to the replacement shoot at its base – see page 46 for details.

# Seasonal Care

Horticultural fleece, hessian or netting draped over the bush or fan in spring will help to protect the blossom from frost damage. Avoid damaging the blossom.

Netting to prevent bird damage is essential for sweet and duke cherries – with unprotected trees, bullfinches devour many buds in winter and early spring – ripe fruits are removed by other birds in summer. Unnetted *Morello* cherries should be harvested at the unripe red-fruit stage to avoid the inevitable attack on ripe fruit.

Water regularly in dry weather – apply up to 30 litres per sq. metre (6 gallons per sq. yard) every 10 days until the dry weather breaks. Avoid heavy watering at regular intervals as fruit splitting can result.

Keep grass away from the trunk. For the first 4 years it is wise to keep the grass about 60–90 cm (2–3 ft) away from the base of the tree.

# Feeding and Mulching

See page 23 – both annual feeding and mulching are necessary for cherries. Acid varieties require more feeding than sweet ones – apply a later spring application of a nitrogen-rich liquid fertilizer.

# Picking and Storage

Sweet cherries should be left on the tree until they are ripe – try one or two in order to tell if the sweet and juicy stage has been reached. Harvest immediately if fruits are starting to crack. Eat as soon as possible after picking, although sweet cherries can be frozen. Red and black ones are best – the yellow ones tend to become discoloured in the deep freeze. Use in fruit salad when thawed.

Use scissors or secateurs when harvesting acid cherries – pulling off fruits can encourage the entry of disease. Cut off with the stalk attached.

# PEACHES AND NECTARINES

Peaches can be successfully grown in England. This delicious fruit is well known in both its shop-bought and canned form; also popular is its smooth-skinned sport, the nectarine. There is no point in pretending that the cultivation of both of these fruits outdoors is anything but a gamble in many areas. In East Anglia and most of the south-east the chance of success is quite high – in the Midlands and the north it is quite low.

Hardiness is not the problem – a cold winter is actually helpful. Trouble may start as early as February when the blossom opens – pollinating insects are scarce and frosty nights are plentiful. High rainfall and high winds are distinctly harmful, and there must be a hot and sunny summer to ensure successful ripening of the fruit.

Plan carefully if you want to take the gamble. Unless the site is extremely favourable, choose a peach and not a nectarine – nectarines are more delicate and yields are much lower. Next, grow as a fan and not as a bush. Bushes are much easier to care for and give higher yields, but fans benefit greatly from the extra warmth and extra frost and wind protection provided by the wall. This wall should face south or south-west – don't even bother to try to grow a peach if it faces in any other direction.

Choose a tried-and-tested variety such as *Peregrine* or *Rochester* – it should have a picking time of mid-September or earlier for southern counties, mid-August or earlier for Midland regions. The earlier the ripening time, the greater the chance of success. Peaches and nectarines are self-fertile, and that means a pollination partner is not necessary. Follow the cultural rules laid down in the following pages, and remember to carry out a peach leaf curl spraying programme every year – see page 56. As you can see from the above information and from the pruning instructions on page 46, there is no easy way to grow peaches and nectarines.

Perhaps the most straightforward technique is to grow *Duke of York* or *Peregrine* on St Julien A or Pixy rootstock in a 38 cm (15 in.) pot. Move the pot indoors from February until the danger of frost is past – then keep on a warm and sheltered patio. After harvest cut back all the branches that have borne fruit to a point where a new shoot has arisen.

## Growing under Glass

Growing a peach or nectarine under glass removes many of the problems which the outdoor tree has to face. The best location is undoubtedly the wall of a lean-to greenhouse – this wall must face south-west, south or south-east. Plant a partly trained fan in well-prepared border soil and then create the basic framework against the supporting wires by following the rules on page 46. Prune the established indoor fan in the same way as an outdoor one.

Close up the house in January, then ventilate if the temperature reaches 16–10°C (60°–65°F). High humidity is essential – spray the floor and mist the foliage in sunny weather. Regular watering is also necessary – in summer that may mean once or even twice a day. Feed regularly with a high-potash fertilizer such as tomato food from the start of flowering until the fruit is almost full-sized. Keep pests and diseases in check – peach leaf curl won't be a problem but red spider mite and greenfly will.

Keep the house cool after harvest. Open up the ventilators and don't worry if winter temperatures fall to freezing point.

## Rootstocks

Peaches and nectarines to be grown as supported fans and free-standing bushes are often grafted on to St Julien A – a semi-vigorous rootstock which produces a plant with a 4.5 m (15 ft) spread. Both Brompton and Peach Seedling are vigorous stocks which are used by professional growers but are not recommended for the home gardener.

Plants grafted on to the semi-dwarfing stock Pixy are more commonly grown as bush, pyramid or fan forms, as compact trees are produced which grow only 2.4–3 m (8–10 ft) high and 3.3 m (11 ft) wide. These dwarfs come into fruit very quickly, but Pixy requires good fertile soil and watering in dry conditions. A more recent introduction from Russia is the dwarfing rootstock VVA-1, which reaches only 1.8–2.4 m (6–8 ft) in height and is suitable for bush, fan and pyramid as well as container growing, so is better suited to the average-sized domestic garden.

## Hand Pollination

Peach and nectarine blossom opens very early in the season when pollinating insects are scarce, so it is a good idea to give nature a helping hand. Dab the flowers gently with a ball of cotton wool or a soft-bristled brush every day or two from the time the buds open until the petals fall. Hand pollination is, of course, essential for peaches and nectarines grown under glass – close the ventilators and dampen the floor afterwards in order to encourage fertilization.

**Duke of York**

**Early Rivers**

**Humbolt**

# Varieties A–Z

## Amsden June

The first peach to ripen – a great advantage in our climate. But *Amsden June* is not a heavy cropper and the fruit is not an attractive colour. Still, the creamy-white flesh inside the round peach does have a good flavour. Grow outdoors or under glass.

**Type:** Peach

**Fruit size:** Medium

**Skin colour:** Greenish-white, flushed with red

Picking time: Mid-July

## Bellegarde

A variety with lots of virtues – the fruit is large or very large, the flesh a deep yellow and the flavour is extremely good. The tree gives high yields but there is a major drawback. It ripens very late and so is only really suitable for the greenhouse.

**Type:** Peach

**Fruit size:** Large

**Skin colour:** Golden-yellow, flushed with dark red

**Picking time:** Mid-September

## Duke of York AGM

If you don't choose *Peregrine* or *Rochester* for reliability then it should be *Duke of York* for fruit quality. The pale yellow flesh is very juicy and the flavour is deliciously refreshing. It crops heavily and does well on a south-facing wall.

**Type:** Peach

**Fruit size:** Large

**Skin colour:** Yellow, heavily flushed with crimson

**Picking time:** Mid-July

## Early Rivers AGM

*Early Rivers* and the rather similar *John Rivers* are the first nectarines to ripen. The yellow flesh is juicy and the flavour is good. Fruit is borne freely, but despite its earliness this variety is not as popular as *Lord Napier*.

**Type:** Nectarine

**Fruit size:** Large

**Skin colour:** Yellow, streaked with red

**Picking time:** Late July

## Fantasia

This produces an excellent crop of vivid orange-red fruits with juicy yellow flesh and a fresh tangy flavour. The tree has a compact habit, making it ideal for small gardens and patio growing. It is easy to grow, with very good resistance to frost.

**Type:** Nectarine

**Fruit size:** Large

**Skin colour:** Orange-red

**Picking time:** Early–mid-August

## Hayles Early

This one crops very heavily and is early enough to make it a sound choice for growing outdoors. Its prolific nature means that thinning is essential in a good season. The flesh is melting rather than firm and the flavour is average.

**Type:** Peach

**Fruit size:** Medium

**Skin colour:** Yellow, flushed and blotched with red

**Picking time:** Late July

## Humboldt

Don't choose this one for outdoors – *Lord Napier* is much more reliable. The situation is different when growing in the greenhouse – *Humboldt* is a very heavy cropper, producing fruit with golden flesh and a distinctive rich flavour.

**Type:** Nectarine

**Fruit size:** Large

**Skin colour:** Orange, flushed with red

**Picking time:** Late August

## Lord Napier AGM

The most popular nectarine for growing at home. It ripens reasonably early and the white flesh is aromatic and full of flavour. It is both a regular and heavy cropper. *Lord Napier* deserves its popularity and is the one to choose for outdoors.

**Type:** Nectarine

**Fruit size:** Large

**Skin colour:** Yellow, flushed with red

**Picking time:** Early August

### Peregrine

**Type:** Peach

**Fruit size:** Large

**Skin colour:** Crimson

**Picking time:** Early August

The number-one peach choice by gardeners, and also the best choice according to some experts. It has the features you want – the flesh is juicy with an excellent flavour and the tree is reliable. Yields are high and nothing succeeds better against a wall outdoors.

### Pineapple

**Type:** Nectarine

**Fruit size:** Medium–large

**Skin colour:** Yellow, flushed with red

**Picking time:** Early September

Early autumn ripening rules out this nectarine for growing outdoors, but it is an excellent choice for greenhouse cultivation. The yellow flesh is melting in texture and the rich flavour is outstanding. A hint of pineapple, according to the catalogues.

### Red Haven

**Type:** Peach

**Fruit size:** Medium

**Skin colour:** Yellow, heavily flushed with dark red

**Picking time:** Mid-August

A round peach with a good reputation for reliability in southern counties. It crops well and the fruit is attractive. The yellow flesh is firm but juicy with a distinct red tinge around the stone. The flavour is good but not outstanding.

### Rochester AGM

**Type:** Peach

**Fruit size:** Medium

**Skin colour:** Yellow, flushed with dark red

**Picking time:** Mid-August

*Rochester* nearly matches *Peregrine* in popularity and is usually considered to be the best all-round variety. The flavour is not as good but the yellow flesh is firm and juicy and the tree is thoroughly reliable. It flowers rather late, thereby missing early frosts.

### Saturn

**Type:** Peach

**Fruit size:** Large

**Skin colour:** Orange-red

**Picking time:** Early August

This is an unusual flat, doughnut-shaped peach, with an orange-red flush on white skin. It has firm white flesh, very sweet with an excellent delicate flavour. This moderately vigorous tree has an upright, spreading habit, but can be susceptible to peach leaf curl.

Peregrine

Rochester

# Planting

A sheltered, sunny site is essential – the soil should be free-draining but also moisture-retentive. November is the best time for a bare-rooted plant, early autumn for a container-grown one. Plant a fan about 23 cm (9 in.) away from the wall – slope the stem slightly inwards when planting.

*Use a strong tie to hold the tree firmly in place.*

| Tree type | Fruit type | Rootstock | Space between trees | Space between rows | Yield per mature tree |
|---|---|---|---|---|---|
| Bush | Peach | Brompton | 6 m (20 ft) | 6 m (20 ft) | 23 kg (50 lb) |
| Bush | Peach | St Julien A | 4.5 m (15 ft) | 4.5 m (15 ft) | 18 kg (40 lb) |
| Bush | Peach | Pixy | 3 m (10 ft) | 3 m (10 ft) | 11 kg (25 lb) |
| Bush | Peach | VVA-1 | 3 m (10 ft) | 2.7 m (9 ft) | 11 kg (25 lb) |
| Fan | Peach | St Julien A | 4.5 m (15 ft) | – | 9 kg (20 lb) |
| Fan | Nectarine | St Julien A | 3.6 m (12 ft) | – | 4.5 kg (10 lb) |
| Fan | Peach | Pixy | 2.7 m (9 ft) | – | 4.5 kg (10 lb) |
| Fan | Peach | VVA-1 | 2.4 m (8 ft) | – | 4.5 kg (10 lb) |

# Pruning

Do not prune in winter and paint all cuts with wound sealant. These measures are necessary to reduce the risk of silver leaf and bacterial canker infection.

## Bushes

Prune in spring. Follow the training procedure for plum bushes – see page 38. Remove all blossom which appears in the season after planting.

For pruning of established trees follow the routine given for acid cherry bushes – see page 42.

## Fans

Buy a partly trained 2- or 3-year-old tree with 8 or more branches. Starting from scratch with a maiden will save a little money, but it will mean a lot of extra work and the loss of a season or two.

### Training a 2- or 3-year-old tree

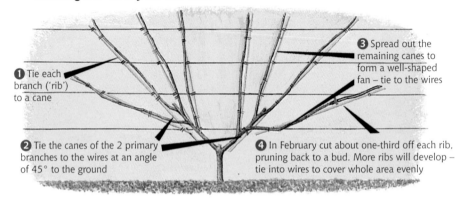

❶ Tie each branch ('rib') to a cane

❷ Tie the canes of the 2 primary branches to the wires at an angle of 45° to the ground

❸ Spread out the remaining canes to form a well-shaped fan – tie to the wires

❹ In February cut about one-third off each rib, pruning back to a bud. More ribs will develop – tie into wires to cover whole area evenly

## Pruning an established tree

In spring remove any shoots growing inwards or outwards. Then pinch out growth buds on each flowering lateral to leave one at the base, one in the middle and one at the top

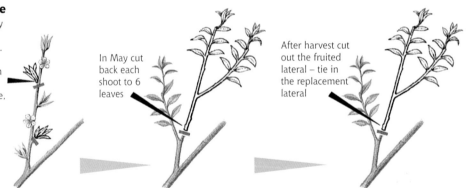

In May cut back each shoot to 6 leaves

After harvest cut out the fruited lateral – tie in the replacement lateral

# Seasonal Care

Hessian or netting placed over the fan or bush in early spring should be used to protect the blossom from frost damage. Remove netting during the day to allow insects to pollinate the flowers.

Netting is also necessary to prevent birds from stripping buds in winter and damaging ripening fruit in summer.

Water regularly in dry weather – apply up to 23 litres per sq. metre (5 gallons per sq. yard) every 10 days until there is prolonged rainfall. Never let the soil around the roots dry out – this is especially important with a newly planted tree close to the house.

Thinning will be necessary if the crop is heavy. Begin when the fruit is marble-sized and continue until it is the size of a walnut. Thin at intervals until there is just one fruit at 15–23 cm (6–9 in.) intervals.

# Feeding and Mulching

See page 23 – peaches require annual feeding and mulching. In addition to the spring application of Growmore, Vitax Q4 or pelleted chicken manure, it is beneficial to use a high-potash feed occasionally during the fruit-swelling stage.

# Picking and Storage

A peach or nectarine is ripe when the flesh around the stalk is soft – the skin also bears a reddish flush. Handle carefully – lift up the fruit in the palm of your hand and twist gently. If ripe it will come away easily. Fruits can be stored in a cool place for about a week – lay them unwrapped in a box lined with tissue.

Peaches can be bottled, dried or frozen – remove the stones before freezing.

# APRICOTS

There are many similarities between apricots and peaches, some of which are obvious. The trees have the same height, spread and general appearance – the fruits of both are stone-hearted, downy-skinned and golden or red with sweet and succulent flesh. Our climate makes their cultivation outdoors a risky business – with both apricots and peaches the standard advice is to grow the tree as a fan against a sunny and protected wall in the south and Midlands, and to grow the tree in a cold greenhouse if you live in Wales or the north.

Frost protection outdoors is essential in most seasons and bird protection is desirable. The blossom is self-fertile so a partner is not required, although the earliness of the flowering season means that hand pollination is recommended. Pot culture is practical – grow *Moorpark* grafted on to St Julien A or Pixy. Begin with a partly trained tree in a 38 cm (15 in.) pot – cut back the main branches by about one half each in early spring until a satisfactory framework has been built up. In subsequent years after harvest cut all branches which have borne fruit back to a side branch. Keep the pot in a greenhouse or conservatory all year round or stand outdoors on a warm patio.

Alongside the many similarities there are several important differences between apricots and peaches. Bacterial canker and silver leaf are not serious threats, so pruning is safe in late winter or early spring. Even more important is the ability of apricots to bear fruit on spurs borne by older wood as well as on 1-year-old wood, like peaches. Because of this pruning is quite a simple task. Soil requirements are also rather different: apricots need more humus and more lime – the ideal locality is in deep and slightly chalky loam.

The choice of varieties is much more restricted than the peach list. Many catalogues just contain *Moorpark*, an apricot which came to England over 200 years ago. It is perhaps the most reliable variety you can choose – pick one of the newer ones if you are adventurous.

## Growing under Glass

Apricots flower even earlier than peaches, so spring frosts can easily ruin the outdoor crop. Even in southern districts you cannot rely on regular fruit production – you get a decent crop of apricots only when the spring is mild and when frost protection is provided.

Growing under glass is much more reliable, and you can choose between cultivating a fan against the south-facing wall of a lean-to or growing it as a dwarf bush in a pot. Pot-grown plants give lower yields than fans, but they are much less trouble.

Any of the varieties described overleaf is suitable. Follow the general cultivation rules set out for peaches on page 43.

## Rootstocks

Apricots are grown on plum rootstocks. Both Brompton and Mussel are too vigorous for garden use – the most popular stock is the semi-vigorous St Julien A which produces a bush about 4.5 m (15 ft) high or a fan which spreads to a similar distance. The semi-dwarfing rootstock Pixy forms a mature bush 2.4–3 m (8–10 ft) in height, while Torinel is also semi-dwarfing and makes a slightly larger plant (3–3.6 m/10–12 ft high), but does better than Pixy on poorer soils. The dwarfing VVA-1 only reaches 1.8–2.4 m (6–8 ft) in height and is suitable for growing in containers and gardens where space is limited.

## Hand Pollination

See page 43.

## Planting

A sheltered sunny site is essential. Good drainage is vital – add well-rotted organic matter if the ground is short of humus. Liming will be necessary if the soil is acid. November is the best time for planting a bare-rooted plant.

| Tree type | Rootstock | Space between trees | Space between rows | Yield per mature tree |
|---|---|---|---|---|
| Bush | St Julien A | 4.5 m (15 ft) | 4.5 m (15 ft) | 18 kg (40 lb) |
| Bush | Pixy | 3 m (10 ft) | 3 m (10 ft) | 11 kg (25 lb) |
| Bush | Torinal | 3.6 m (12 ft) | 4.5 m (15 ft) | 11 kg (25 lb) |
| Fan | St Julien A | 4.5 m (15 ft) | – | 9 kg (20 lb) |
| Fan | Pixy | 2.7 m (9 ft) | – | 4.5 kg (10 lb) |
| Fan | VVA-1 | 2.4 m (8 ft) | – | 4.5 kg (10 lb) |

Moorpark

Tomcot

# Varieties A–Z

### Goldcot

This is a modern, cold-hardy apricot, which is better suited to the UK climate than most. It is reliable and crops heavily. If you are keen to try growing apricots in the garden, this is probably your best choice. The flavour for eating fresh is only average, but it is excellent for cooking.

**Type:** Apricot
**Fruit size:** Medium–large
**Skin colour:** Golden-yellow, flushed orange
**Picking time:** Mid-August

### Golden Glow

This reliable apricot regularly produces good crops of juicy, succulent fruits. It crops well as a free-standing tree or as a trained tree in UK conditions. The tree is very hardy, and has good resistance to bacterial canker, making it a good choice for wet regions.

**Type:** Apricot
**Fruit size:** Medium
**Skin colour:** Orange-yellow
**Picking time:** Early August

### Moorpark AGM

By far the most popular variety and the only one offered by many nurseries. The orange flesh of this round variety has a very good flavour and the tree is renowned for its reliability – but it is prone to die-back.

**Type:** Apricot
**Fruit size:** Large
**Skin colour:** Golden-yellow, flushed with red
**Picking time:** Late August

### New Large Early

The oval fruit, with a very fine delicious flavour, is larger and ripens earlier than *Moorpark*. A tree of moderate vigour but very productive and hardy, with good die-back resistance. Favoured as an especially good variety for bottling.

**Type:** Apricot
**Fruit size:** Large
**Skin colour:** Golden-yellow, flushed with red
**Picking time:** Early August

### Tomcot

A new apricot from France, which produces excellent yields of extremely large, juicy fruits. The orange flesh is firm with an excellent flavour. For the best results, fan train against a south-facing wall. This variety is a suitable choice for northern gardens.

**Type:** Apricot
**Fruit size:** Large
**Skin colour:** Golden-orange, flushed with red
**Picking time:** Mid–late July

# Pruning

### Bushes

Plant a 2- or 3-year-old partly trained tree. Train as for plum bushes – see page 38. Prune in early February.

For pruning of established trees follow the routine given for acid cherry bushes – see page 42.

### Fans

Plant a 2- or 3-year-old partly trained tree. Follow the training procedure for peach fans – see page 46.

For pruning of established trees follow the routine given for plum fans – see page 38.

## Seasonal Care

Hessian or netting placed over the fan or dwarf bush in early spring should be used to protect the blossom from frost damage. Support the net away from the flowers to avoid harming them. Remove during the day.

Water regularly in dry weather – apply up to 15 litres per sq. metre (3 gallons per sq. yard) every 10 days until there is prolonged rainfall. Never let the soil dry out – this is especially important with a newly planted tree and when the fruits are beginning to swell.

Thinning will be necessary if the crop is heavy. Wait until the average fruit is about the size of a cherry and then remove the smallest and most crowded ones. Repeat the process until the fruits are 5–8 cm (2–3 in.) apart.

## Feeding and Mulching

See page 23. Each spring mulch the plants with a 2–3 cm (1 in.) layer of organic matter.

## Picking and Storage

Apricots become soft several days before they are ready for picking. Harvest them when they come away easily from the tree. Handle gently – the fruit bruises very easily. Keep indoors for about a day before eating. Apricots store much better than peaches – they can be kept for several weeks in a cool place. Lay them unwrapped in a box lined with tissue.

Apricots can be bottled, dried or frozen – remove the stones before freezing.

# FIGS

Figs are a strange fruit – their cultivation involves techniques which you will find nowhere else in this book. It is not even easy to classify them. They are treated as tree fruit in this and several other textbooks, but as soft fruit in some catalogues.

Figs are quite easy to grow. They are not grafted on to rootstocks and they do not need a pollination partner. Any soil will do as long as it drains reasonably freely, and the trees can withstand the winters in southern and western counties. Spraying is not necessary as pest and disease attacks are rare. This ease, however, only applies to the tree as an ornamental plant – it is difficult to induce a fig tree to produce a worthwhile crop every year. Read this section before deciding to buy one – like most gardeners you will probably decide that a peach, grapevine or melon would be a much more desirable exotic.

The first problem concerns root growth – if left unrestricted the tree grows too vigorously and few fruits develop. To prevent this, the roots are restricted by planting in a 38 cm (15 in.) pot or a lined pit – see below. Pot-grown figs can be trained as a fan against a wall, as a standard on a clear stem 1–1.5 m (3–5 ft) high, or maintained as a dwarf bush. Bush figs can be kept in a greenhouse or left outdoors for most of the year, then brought indoors during the leafless period when severe frosts threaten.

A sheltered wall facing south or south-west is the place for a fig. Greenhouse culture is feasible – you will get two crops instead of one each year if you can maintain a temperature of 12–13°C (55°F) from January onwards. But do think carefully – figs are leafy and cast a lot of shade.

Another problem is winter protection for outdoor crops. This year's figs were present last year as tiny (embryo) fruits. These embryo fruits and young shoots must be individually covered during winter.

During the growing season the crop needs regular attention. Frequent watering is essential – during dry weather in summer you will have to irrigate twice a week. Shoots have to be pruned in June and unwanted figs have to be removed in late September – see page 50. Obviously figs are *not* for the lazy gardener, but they are a worthy challenge for the keen gardener who wants to see just how delicious a home-grown fig can be. There are two types – the green varieties which have greenish skin and pale flesh, and the purple varieties with brown or purple skin and red flesh. For outdoor cultivation the reliable *Brown Turkey* is the usual choice – under glass the superb-tasting *Bourjasotte Grise* is worth trying.

# Planting

Choose a well-sheltered spot which is exposed to the sun. Plant fans 3.6–4.5 m (12–15 ft) apart. November to March is the recommended planting period – March is the best time.

The traditional method is to construct a fig pit, as shown here. An easier method is to plant in a large tub or a 38 cm (15 in.) pot. Make sure that the container has adequate drainage holes, crock well and use soil-based compost. In both pits and pots you should plant very firmly.

The pot can be stood in a sunny sheltered part of the garden, to be moved into a shed when frost threatens. If the fig is to be grown against a wall as a fan then the pot should be sunk up to its rim in the earth close to the wall.

Repotting will be necessary every 2–3 years – do this in late winter. Alternatively, remove the top 5–8 cm (2–3 in.) of existing compost and replace with fresh compost containing a controlled-release fertilizer in late winter or early spring.

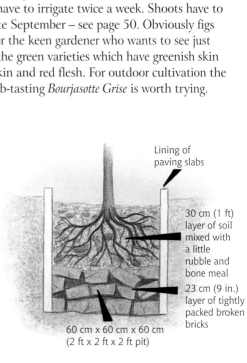

Lining of paving slabs

30 cm (1 ft) layer of soil mixed with a little rubble and bone meal

23 cm (9 in.) layer of tightly packed broken bricks

60 cm x 60 cm x 60 cm (2 ft x 2 ft x 2 ft pit)

# Varieties A–Z

### Bourjasotte Grise

This one is strictly for growing in the greenhouse – it is too tender for outdoors. The fruits are rounder than the other varieties listed here and the flavour is better – probably the richest and sweetest fig you can grow. Good for pot cultivation.

**Type:** Green fig
**Fruit size:** Medium–large
**Skin colour:** Pale green
**Picking time:** September

### Brown Turkey

For many years this has been the number-one choice – it is suitable for outdoors or under glass. A reliable and heavy cropper producing oval fruit – the red flesh has a rich and sweet flavour. Often the only fig offered in the catalogue.

**Type:** Purple fig
**Fruit size:** Medium–large
**Skin colour:** Brownish-red. Blue bloom
**Picking time:** August–September

### Brunswick

Not as popular as *Brown Turkey*, but it does come next in the rankings. An oval-shaped fruit like *Brown Turkey*, but it is larger and also ripens about a couple of weeks earlier. The flavour is very good but the tree is not as reliable as its rival and the yields are lower.

**Type:** Green fig
**Fruit size:** Very large
**Skin colour:** Greenish-yellow, flushed with brown
**Picking time:** Mid-August

### Panache

This variety has flattish to oval-shaped fruits which take on a violet hue in cool weather. The red flesh is aromatic, with a sweet, rich flavour. Ideal for pot culture in a greenhouse or conservatory as it does not perform well outside in the UK unless in a very warm spot.

**Type:** Green fig
**Fruit size:** Medium
**Skin colour:** Pale green, striped yellow
**Picking time:** September

### White Marseilles

This variety ripens at about the same time as *Brown Turkey*, but that is the only similarity. The fruit of *White Marseilles* is distinctly pear-shaped, and the whitish flesh is almost transparent. It has a good reputation for pot culture.

**Type:** Green fig
**Fruit size:** Large
**Skin colour:** Pale green
**Picking time:** August–September

Brown Turkey

White Marseilles

# Pruning

Buy a 2- or 3-year-old partly trained tree. For a bush follow the training procedure for apple bushes – see page 24. For a fan follow the training procedure for peach fans – see page 46.

Pruning of established bushes and fans takes place in June. All young shoots should be pinched back so that only 5 leaves remain – this will encourage new fruiting shoots to form. In the case of fans tie these shortened growths to the wires.

Thin out the fruits in late September. The embryo figs which should be retained are close to the end of the shoots and are about pea-sized. These will develop into fruit for picking next year. At this late-September stage all the ripe fruit will have been picked, but you will also find some cherry-sized figs which will not ripen – remove them.

In spring clean up the tree. Remove branches killed by frost and on a fan remove buds which are pointing directly inwards or outwards from the wall.

# Seasonal Care

Winter protection of young shoots and embryo fruits is vital. Move a pot-grown plant into a shed, garage, porch, conservatory or greenhouse. For plants which can't be moved indoors it will be necessary to tie a covering of straw loosely around the tender parts, or cover them with a layer of horticultural fleece, held in place with clothes pegs. Put on this cover in late autumn – remove gradually during April and May. Protect the roots of pot-grown plants by wrapping several layers of plastic bubble wrap around the container.

Regular watering is essential as the tree has restricted root growth. The root ball must never be allowed to dry out – adequate moisture is especially important when the fruits are swelling in early summer. Watering can be reduced as harvesting approaches.

In late summer it is usually necessary to hang a net over the tree to protect the ripening fruit from birds.

# Feeding and Mulching

Do not overfeed. Apply a light dressing of Growmore, Vitax Q4 or pelleted chicken manure in spring and spread an organic mulch over the soil surface. When the fruits start to swell apply a high-potash liquid fertilizer, such as tomato feed.

# Picking and Storage

A fig is ready for harvest when the stalk weakens and the fully coloured fruit hangs downwards. There will be other tell-tale signs – the skin may be cracked or there may be a drop of nectar at the base. Gather the fruit carefully – it will keep in a cool place for several weeks. Figs can be dried in the airing cupboard. Turn daily – drying will take 4–7 days.

# MEDLAR

An ornamental tree, with a spreading habit, pretty late-spring to early-summer blossom and good autumn colour. The medlar tree prefers a sunny sheltered spot but will tolerate partial shade and most soils. The fruit is harvested in October before the first frosts. It is russet brown and the size of a small apple, and the seeds are visible through the gaping cup below the calyx. Even odder is the preparation for serving. After the fruits are picked, they are then allowed to decay ('blet') for several weeks. The flesh turns brown, soft and sweet – it is scooped out with a spoon.

# MULBERRIES

Mulberries have been grown in Britain for hundreds of years and they are worth considering in southern areas – they are disappointing in shady northern gardens and in heavy, water-retentive soil. The place for a mulberry tree is in a large lawn where its gnarled branches and stately head can be admired. It is remarkably trouble-free and the flowers are self-fertile, but bear in mind that the eventual height and spread will be about 6 m (20 ft); growth is slow in the early years and the tree is slow to produce leaves in the spring. Worst of all, you will have to wait about 7 years before you can taste the fruit.

Buy a 3–5-year-old trained tree and plant in March – a *Black Mulberry* is recommended. Stake for the first few years and prune as little as possible – the branches bleed when cut. The large dark red or black raspberry-like fruits have a sweetish but tart flavour and are harvested in August or September. Mulberries can be eaten fresh or used for jam, jellies or wine.

# QUINCES

Even if you have no use for the fruit it is worth growing a quince tree. The gnarled trunk, grey bark and leathery dark green leaves give it a distinctly oriental look, and in June attractive white or pale pink blossoms appear, 4–5 cm (1½ – 2 in.) in diameter, followed by aromatic apple- or pear-shaped fruit. This is a plant for the south and the Midlands. It prefers moist soil and you will need space for a tree that grows to about 3.6 m (12 ft).

Plant between November and March, or at any time if it is container-grown. Choose a 2–4-year-old trained plant – *Vranja* (large, golden, pear-shaped) is the popular variety, but you may also find *Champion* (large, apple-shaped) and *Portugal* (large, early, orange, pear-shaped). During the first few years cut back the main leaders in winter to remove about half of the previous year's growth. Keep the centre of the bush open. Once the mature framework is established, prune as little as possible – simply remove dead, dying, diseased or crossing branches in winter. Pick the fruits in October before the first frosts arrive. Store them in trays in a cool, dark place until they change from green to yellow. This will take 4–8 weeks. Quinces are usually too tart and gritty to be eaten raw but can be poached and make delicious jellies.

# The Major Pests and Diseases

## APPLES

The major pests include aphids (especially on young trees) and caterpillars (especially winter moth). Codling moth is the main danger to developing fruit – in addition keep watch for apple sawfly and capsid. Scab and powdery mildew are the diseases you are most likely to see but you should also look out for canker on stems and branches and brown rot on fruits.

*Apple powdery mildew.*

## PLUMS

Aphids are a major pest as with other tree fruit. Birds can seriously damage the flower buds and wasps, plum moth maggots and plum sawfly caterpillars can ruin the fruit. Watch for red spider mite in hot settled weather and look for the tell-tale signs of silver leaf, brown rot and bacterial canker.

## PEARS

As with apples, both aphids and caterpillars are major pests. Pear and cherry slugworm can also be a serious nuisance and pear midge is an important cause of premature fruit drop. Among the major diseases are scab and the more deadly but less often seen fireblight.

## CHERRIES

Aphids (cherry blackfly) can be devastating. Birds are another serious problem, attacking both flower buds and ripening fruit. As with plums, both silver leaf and brown rot are important diseases and so is bacterial canker.

## PEACHES

Aphids and red spider mite are major pests when the weather is hot and dry. There are several important diseases, including peach leaf curl, silver leaf, bacterial canker and split stone.

### Stages of apple and pear development

**Bud swelling**

**Bud break**

**Bud burst**

**Mouse ear**

**Green cluster**

**Pink bud**
(apples)

**White bud**
(pears)

**Blossom time**
Early to mid-May
(apples)
Late April to early
May (pears)

**Petal fall**
When nearly
all petals
have fallen

**Fruitlet**
Mid-June

**Fruitlet**
Early July

# HOLES IN LEAVES

Holes in the foliage of fruit trees are a common sight. The culprit in spring is usually the winter moth or mottled umber moth – during late spring and summer there are a number of pests which can attack fruit trees in this way. The common causes of torn or skeletonized leaves are shown on this page, but some caterpillars which attack soft fruit (yellowtail moth, magpie moth, etc. – see page 103) occasionally attack apples, pears and plums.

## Shot hole disease

Either a bacterial or fungal disease which attacks plums, peaches and cherries. Brown spots appear on the leaves which turn into small holes. No other symptoms appear and trees are not seriously affected. As only weak trees are attacked, increase vigour by mulching and feeding in spring. Drought makes the trees susceptible, so water in dry weather.

## Bacterial canker

A serious disease of plums, cherries and other stone fruit. Pale-edged spots are the first sign; later, gum oozes from the bark and affected branches die back. Act quickly to save the tree. Cut out diseased branches and paint cuts with wound sealant. No spray reliably gives complete control, but copper oxychloride and/or copper sulphate and hydrated lime can check the disease if applied in the early stages.

## Capsid bug

The first signs of damage on apple leaves are reddish-brown spots. As the leaf expands these tear to give ragged brown-edged holes, usually puckered and distorted. Spraying is not usually necessary but deltamethrin, lambda-cyhalothrin or thiacloprid applied at petal fall should protect fruit.

6 mm (¼ in.) active bugs

## Pear and cherry slugworm

These slimy slug-like insects attack pears and cherries in June–October. They feed on the upper surface and foliage may be skeletonized. Spray with a contact oil-based or pyrethrin-based insecticide or deltamethrin, or lambda-cyhalothrin if leaves are seriously affected.

1 cm (½ in.) black grubs

## Fruit tree tortrix moth

Leaves are spun together or a leaf is attached to the fruit. The small active caterpillars feed within this protective cover. If touched on the head they wriggle backwards. Hand-pick spun leaves if practical. Acetamiprid or thiacloprid will control them, but a thorough drenching spray is necessary.

## Apple leaf skeletonizer

Occasional pest of apples, pears and cherries in south-east England. The 1 cm (½ in.) caterpillars are yellow with black spots, and their feeding skeletonizes the foliage. Most active in late summer, when damaged leaves should be picked and burnt. This pest is rarely serious enough to justify an insecticide.

## Tent caterpillar

These caterpillars produce a tent of silken webs, within which they feed on the foliage. The tents should be picked off and burnt if practical. Alternatively spray with acetamiprid, thiacloprid, lambda-cyhalothrin or deltamethrin as soon as tents start to appear.

Lackey moth

Browntail moth

Small ermine moth

## Caterpillar

**Mottled umber moth**
Smooth 'looper' caterpillar, which attacks trees in spring.

**Winter moth** Green 'looper' caterpillar which devours young leaves and may spin them together. They eat the new leaves in spring and later feed on the petals and flower stalks. Encircle each trunk with a greaseband barrier from September to March. If caterpillars are seen, spray with acetamiprid, thiacloprid, lambda-cyhalothrin or deltamethrin.

**Vapourer moth**
Colourful caterpillar that eats leaves in May–August. Spray with acetamiprid, thiacloprid, lambda-cyhalothrin, deltamethrin or a contact oil-based or pyrethrin-based insecticide.

## Vine weevil

A serious pest on container-grown fruit and strawberries. The adult females (rarely seen) eat semi-circular notches along the margins of leaves, then lay eggs at the base of the plant. These develop into small C-shaped white grubs with a black or brown head, which feed on the roots of a wide range of plants. Control with parasitic eelworms watered around the base of affected plants.

# LEAF AND SHOOT PROBLEMS

Apart from holes (page 53), all sorts of problems can occur on the foliage. Leaves may be twisted, discoloured, blistered or covered with mould. Do not waste time and money spraying against everything – just worry about the serious pests and diseases. These are blackfly on cherries, greenfly on young trees, silver leaf on plums, scab and mildew on apples and pears, leaf curl on peaches and die-back on apricots.

### Apple twig cutter

Blue weevil, 4–5 mm (⅙ in.) long, bites through the young shoots of apple, pear or plum after egg-laying in June. These shoots wither and may remain hanging on the tree. Small trees are most at risk, and can be protected by spraying with pyrethrins at white or pink bud. A minor pest – treatment is rarely necessary.

### Clay-coloured weevil

Brown weevil, 5–6 mm (¼ in.) long, gnaws the bark of woody stems. Young apple trees are sometimes crippled by this night-feeding pest. Control is not easy – hoe the surrounding soil and spray both the tree and surrounding ground with acetamiprid or thiacloprid. Alternatively, encircle each trunk with a greaseband barrier.

### Die-back

Begins at the shoot tips and progresses slowly downwards. More common in stone fruit than apples or pears. There are a number of reasons, including diseases such as canker. If no disease is present, waterlogging is a likely cause. Cut out dead wood. Mulch trees. Improve drainage.

### Leopard moth

Large spotted caterpillar, about 5 cm (2 in.) long, which burrows for 2–3 years inside the affected branch. Leaves wilt – the branch may die. If the trunk is attacked the tree can be killed. This pest is difficult to control by chemical means – it is often necessary to cut out and burn the affected branch.

### Frost

The effect of frost on the foliage becomes apparent as the leaves expand. The lower surface becomes puckered and blistered, and the skin over the blister cracks, exposing the tissue beneath. The leaves around the flower trusses are the worst affected. The leaves do not fall and there is no treatment.

### Woolly aphid

Colonies of aphids live on stems and branches, secreting white waxy 'wool' which protects them. Their presence does little direct harm, but the corky galls they cause are a common entry point for canker spores. Spray with acetamiprid or thiacloprid, preferably while the affected tree is in leaf.

### Leaf midge

A common pest of pears; less common on apples. The grubs cause the edges of the leaves to roll upwards. Young foliage is more likely to be attacked than old leaves. Affected leaves may turn red and fall. Control is difficult as the grubs are protected from sprays. Pick off and burn affected leaves.

### Leaf hopper

Pale mottled patches appear on leaves, caused by small, lively greenish insects feeding on the under-surface. Empty white skins are a tell-tale sign. They can be controlled by spraying but this is only necessary in the case of a bad attack. Contact sprays will not do – a systemic insecticide such as acetamiprid or thiacloprid is necessary.

## Nutrient deficiency

An abnormal change in leaf colour often indicates a shortage of an essential element. If the symptoms of iron or magnesium deficiency are severe, water on the soil around the trunk with a chelated product. Feed and mulch the tree in spring. Make sure the soil pH is suitable before planting.

**Nitrogen shortage**
*Red and yellow tints*

**Potash shortage**
*Leaf edge scorch*

**Magnesium shortage**
*Brown between veins*

**Iron shortage**
*Yellow between veins*

## Silver leaf

The most serious disease of plums, which can also attack apples, cherries and peaches. The spores enter through a wound or pruning cut, and the first sign is silvering of the leaves. Die-back of shoots occurs; wood is stained. Cut out dead branches 15 cm (6 in.) below level of infection (a point where staining of silver leaf disease is absent). Paint cuts with a wound paint. Dig out tree if bracket-like toadstools have appeared on the trunk.

*Cross-section of shoot*

## False silver leaf

A common disorder which looks like silver leaf at first glance. Leaves are silvery, but the effect appears all over the tree rather than progressively along a branch. A cut branch reveals that the staining of silver leaf disease is absent. The cause of false silver leaf is starvation or irregular watering. Feed regularly and put down an organic mulch in spring.

## Leaf blister mite

This microscopic mite can be a serious pest of pears grown against walls. Leaves are spotted with yellow or red blisters which later turn black. Affected leaves fall early, and fruit may be blistered. Pick off and burn diseased leaves as soon as they are noticed, or spray before leaf fall with a contact insecticide based on plant oils, fish oils or fatty acids.

## Greenfly

Several species of greenfly can attack apples, pears, peaches and plums – in some seasons they are a serious pest. Some greenfly cause yellowing and curling of the young leaves. The rosy leaf-curling aphid causes reddening and distortion of apple foliage. Greenfly secrete sticky honeydew and stunt shoot growth. Spray with thiacloprid or insecticidal soap at bud burst. Repeat at green cluster.

## Red spider mite

The first sign of red spider mite attack is a faint mottling of the upper leaf surface. In warm weather a severe infestation may occur – the leaves turn a bronze colour, become brittle and die. Examine the underside of the foliage with a magnifying glass for the tiny mites. The fruit tree red spider mite is the common species; on pears you may find the Bryobia mite. Spray with acetamiprid in late spring or early summer. Or with plant oils, plant extracts or fatty acids and repeat if necessary 3 weeks later.

*Fruit tree red spider mite*

*Bryobia mite*

## Cherry blackfly

Only one aphid, the cherry blackfly, attacks cherries, but its effect can be devastating. Leaves may be severely curled and shoot growth halted. It is therefore essential to kill these pests before they become established. Spray with acetamiprid, thiacloprid, lambda-cyhalothrin, deltamethrin or a contact oil-based or pyrethrin-based insecticide when they are seen.

## Scab

A serious disease of apples and pears, which can attack all parts of the tree. Leaves bear dark green or brown spots; twigs are blistered and fruits are badly disfigured (page 60). Attacks are worst in warm, damp weather. Spray with myclobutanil at green cluster, pink or white bud, petal fall and 14 days after petal fall. Rake up and burn fallen leaves in autumn; prune scabby shoots in winter.

## Mildew

Young leaves, shoots and flower trusses of apples and pears may appear grey in spring due to infection of this white powdery mould. It is a serious disease – growth is stunted, diseased flowers do not set and leaves may fall. *Cox's Orange Pippin* is very susceptible. Prune and burn all infected twigs. Use a spray containing myclobutanil or potassium phosphate, or a spray based on fish oil – repeat applications will be necessary.

## Peach leaf curl

Large reddish blisters develop on the leaves. Apart from making the tree unsightly, this serious disease of outdoor peaches leads to early leaf fall and weakening of the tree. The fungus over-winters in the bark and between bud scales, not on fallen leaves. Spray with copper in mid-February, 14 days later and again just before leaf fall. Grow a resistant variety.

## Pear rust

A fungal disease that causes bright orange spots on the upper surfaces of pear leaves and brown, gall-like outgrowths on the corresponding underside in summer and early autumn. Remove and destroy infected leaves as soon as they are noticed. There are no approved products available to control this disease.

## Plum rust

This fungus shows as dusty orange spots on the upper surface of the leaf and dark brown markings on the underside. In severe attacks these can be very close together with many spots on each leaf. There is normally no need to control this disease, because it does little damage and the fruits are not affected.

## Apple leaf miner

This is a fairly common pest, frequently found on apples and sometimes on cherries. As the larvae feed within the leaf they create lines of dead cells (mines). If several mines develop, the tissue becomes discoloured and the affected leaf may gradually shrivel and die while remaining on the branch. Spray with thiacloprid or acetamiprid.

# BARK AND TRUNK PROBLEMS

Some problems of fruit tree trunks, such as scale and crown gall, are surface disorders which can be readily removed. Others, such as bacterial canker, are deep-seated and extremely serious. A few can prove fatal – potential killers include fireblight on pears, honey fungus on apples and canker on most tree fruits. So keep a watch for the troubles shown here.

## Bacterial canker

Cankers are flat and may not be easily noticed, but the effect on stone fruit is serious. Attacked branches produce few leaves and soon die. Gum oozes from the cankers. For leaf symptoms see page 53. Cut out diseased branches; apply a wound paint. Spray trees with a copper-based fungicide in August, September and October.

## Crown gall

Large, brown and warty outgrowths sometimes occur on the stems and roots of fruit trees. Attacks are usually restricted to young trees on badly drained sites. Mature trees are not harmed. Younger ones may lose vigour – cut off galls and paint the affected area with wound paint in autumn.

## Coral spot

Raised pink spots appear on the surface of affected branches. Dead wood is the breeding ground for the fungus, and the air-borne spores infect living trees through cuts and wounds. Never leave dead wood lying about. Cut out all dead and diseased branches and always paint pruning cuts with a wound paint to protect the cut surfaces.

## Canker

The bark shrinks and cracks in concentric rings. A tell-tale sign is the presence of red growths in winter. A serious disease of apples and pears, especially on badly drained soil. Cut off damaged twigs – cut out canker from stems and branches. The branch will be killed if the canker encircles it. Paint with a wound paint.

## Scale

Scale insects are a pest of neglected trees. These tiny shell-like creatures do not move. Uncommon outdoors, but with greenhouse fruit they may completely cover the branches with a scaly coat. All types of tree fruit can be affected. Usually ignored, but with a severe attack spray with acetamiprid, deltamethrin, lambda-cyhalothrin or plant oils/extracts in early July when the more vulnerable newly hatched scale nymphs are present. Prune heavily infested shoots, and a winter wash can be used in December–January when the plants are fully dormant.

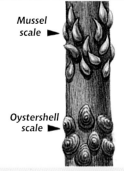

*Mussel scale* ▶

*Oystershell scale* ▶

## Gummosis

A disorder of cherries and plums – usually occurs after freezing weather. Patches of gum appear on the surface of branches and trunks. This gum arises from healthy wood – with the much more serious bacterial canker (left) the gum oozes from diseased tissue. Mulch and feed tree in spring.

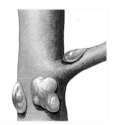

## Fireblight

A devastating disease of pears which can occur on apples. Affected shoots wilt and die. Old cankers ooze in spring. Tell-tale sign is the presence of brown, withered leaves which do not fall. Cut out all diseased wood to 60 cm (2 ft) below affected area (a point where reddish-brown staining under the bark is absent). Disinfect tools. The tree is killed once the disease spreads to the trunk.

## Honey fungus

Honey fungus (root rot, armillaria disease) is a common cause of the death of apple trees. A white fan of fungal growth occurs below the bark near ground level. On roots, black 'bootlaces' are found. Toadstools appear in autumn at the tree base. Burn stems and roots of diseased trees. Currently no fungicide is available that is effective against this disease.

## Papery bark

The bark becomes thin and peels off in sheets. Die-back occurs if a shoot is girdled. This form of bark canker affects apple and pear trees in poor condition, usually due to bad drainage. Remove loose bark – paint exposed surface with a wound paint. Improve drainage. There is no cure – the only treatment is to mulch and feed in spring.

## Split bark

A crack may appear in the bark at any time of the year. The cause can be a severe frost, but the usual culprit is a cultural problem such as poor drainage or inadequate staking. Another possible cause is heavy rain after drought. Cut away any dead wood – apply a wound paint to keep out disease. Feed and mulch the tree to restore good health.

## Collar rot

Collar rot is a disease of the scion which usually only attacks mature trees over 10 years old (and mainly *Cox*). This soil-borne disease first shows as a wet sunken patch on the trunk which has a distinct orange/red-brown rot under the bark. Control by cutting away the sunken patches and treating the wounds with a copper-based spray.

## Sucker

The developing buds of apples and pears are attacked by young suckers. The blossom trusses turn brown as if attacked by frost, but careful examination reveals yellowish-white insects which look like flattened aphids. Another tell-tale sign is the presence of  sticky honeydew. Spray with a contact insecticide, such as deltamethrin, lambda-cyhalothrin or plant oils/extracts at the green cluster stage.

## Cherry fruit moth

These green caterpillars can be a serious pest in southern counties. They feed inside the flower buds where they eat both petals and stamens. They continue to feed on the open flowers and then inside the young fruitlets. They drop to the ground in late May. It is not usually necessary to spray against this pest, but if attacks have been heavy in the past apply acetamiprid or thiacloprid at bud burst and again at white bud.

## Blossom wilt

In a mild, wet spring this disease can be extremely serious on plums, when the tree appears to have been scorched. Apples and pears can also be badly affected. The blossom trusses wilt and turn brown – shoots are killed in a bad attack. Remove infected blossoms and dead twigs in spring. In summer remove all fruit affected by brown rot (page 60). Apply a copper-based fungicide spray just before flowering and again 14 days later.

## Frost

Frost at blossom time causes the flowers to turn brown and drop off. This can, of course, be serious if little blossom has been produced, but in a year of abundant blossom the thinning effect of a slight frost is beneficial. If your garden is on a sloping site, open part of the lower boundary to air movement or you will create a 'frost pocket'. Choose a late-flowering variety.

## Birds

Birds can be extremely destructive to flower buds, the main culprits being bullfinches and sparrows. All tree fruit can be damaged, with cherries, plums and pears the worst affected. The outer bud scales are pushed aside and the central portion of each bud pecked out during late winter and spring. Unfortunately, spray repellents may not be effective as the outer sprayed portion of the bud is not eaten. The best answer is plastic netting if this is practical.

## Blossom drop

It is annoying to discover that a tree which was full of blossom has lost its flowers without setting fruit. To find the cause you must know whether this trouble takes place year after year or is an unusual occurrence. If it is an annual event then the most likely cause is the absence of a pollination partner. Most trees will not set fruit unless another variety with compatible pollen is growing nearby. If blossom drop does not usually occur, then poor weather at flowering time is the most probable reason. Frosting of the blooms is a common problem on cold, exposed plots. Very dry air can result in poor pollination and a wet, cold spring reduces the activity of pollinating insects.

## Apple blossom weevil

A common and troublesome pest of apples, which also occasionally affects pears. In spring the grubs feed inside developing flower buds. The stalks remain green and healthy, but the petals turn brown and fail to open. If one of these 'capped' blossoms is cut in two, a white grub or brown beetle will be found inside. This is the apple blossom weevil or the closely related and much less common pest, the apple bud weevil.

Apple blossom
weevil:
'V' on back

Apple bud
weevil:
no 'V' on back

*Spray with thiacloprid at bud burst.*

## Winter moth

The winter moth caterpillar is a serious pest in early spring, devouring the new leaves. When the flowers are fully open the caterpillars often feed on the petals and flower stalks. The 3–4 cm (1½ in.) 'looper' caterpillars may even attack young fruitlets, biting holes in the surface. Protect trees by encircling each trunk with a greaseband barrier from September to March. If numerous caterpillars are seen on the leaves, spray with acetamiprid, lambda-cyhalothrin or deltamethrin.

# FRUIT PROBLEMS

It is unfortunate that so many troubles can occur after the fruit has set. The most likely problems are illustrated below, but others can take their toll. Birds can be a pest, often stealing the whole crop from a cherry tree or picking holes in ripening apples. Greenfly attacks can result in small distorted fruit, and frost can produce corky stripes on the outside and brown watery flesh within.

## Split stone

The outside of an affected peach is holed at the stalk and inside the stone is split and rotten. The usual cause is poor pollination or irregular watering – next year hand pollinate the flowers and spray them gently in dry weather. Keep the soil moist and apply an organic mulch. Apply a dressing of lime if the soil is acid.

## Apple fruit rhynchites

Brown weevil, 5 mm (⅕ in.) long, creates numerous small holes with its snout over the surface of developing fruit. Attacks cease in July, but affected fruits remain disfigured and distorted as they enlarge. The flesh remains sound. This pest is controlled by sprays used against major fruit pests.

## Wasp

All types of tree fruit may be attacked by wasps. Soft-skinned types such as plums are favoured and may be devoured; apples and pears usually escape if the surface is undamaged. Spraying is not effective – the only answer is to find and destroy the nest with a proprietary wasp killer. Do this job at dusk to avoid harming beneficial insects.

## Earwig

All types of tree fruit, especially the soft-skinned types, may be attacked by earwigs. Apples are attacked in late summer, the surface being holed and the flesh below eaten out. The ground below the trees should be well cultivated – inverted flower pots filled with straw treated with pyrethrin-based dust should be placed on canes.

## Chafer beetle

The 6–7 mm (¼ in.) brown-coloured cock-chafer and the 10–12mm (½ in.) bluish-green garden chafer are responsible for biting out holes in small apple fruitlets in May and June. Badly bitten fruits are useless, but spraying is not practical as pests will have left when damage is seen.

## Pear midge

Attacked pear fruitlets become deformed and blackened, usually falling from the tree. Cut fruit reveals a large central cavity and numerous tiny grubs. Pick and burn blackened fruit, and keep the soil well cultivated. If it was a problem last year, spray with thiacloprid, acetamiprid, lambda-cyhalothrin or deltamethrin at white bud.

## Plum sawfly

The tell-tale sign of this serious pest of plums is a hole surrounded by sticky black 'frass'. Inside will be found the 10–12mm (½ in.) creamy-white grub of the plum sawfly. Damaged plums fall to the ground before maturity. Cultivate the soil around the trees and spray with thiacloprid, acetamiprid, lambda-cyhalothrin or deltamethrin 7–10 days after petal fall.

## Pear stony pit

This is a serious virus disease of pears – affected trees cannot be cured and have to be dug out and destroyed. Diseased fruit is small and misshapen with the surface covered with dimples and lumps. The flesh is woody and inedible. The disease usually occurs on old trees and can spread to surrounding trees.

## Storage rot

Most apple rotting in store is due to bitter rot (gleosporium). If a large quantity of fruit is to be stored, take preventative measures at once. There are no longer any chemical sprays for treating trees before picking. Remove and destroy diseased fruit immediately and do not prune before February. Check stored fruit – destroy rotten ones at once.

## Tortrix moth

The caterpillar of the fruit tree tortrix moth spins a leaf on to the fruit and feeds underneath this cover. The summer fruit tortrix also spins on a leaf cover, and removes extensive areas of skin and surface flesh. These caterpillars can be controlled by spraying with thiacloprid, acetamiprid, lambda-cyhalothrin or deltamethrin in mid-June and applying a repeat spray later if required.

## Scab

This serious disease affects apples and pears. Dark-coloured scabs appear on young fruit – as the fruits grow, large corky areas develop. The surface may be cracked but the flesh remains sound. Attacks are worst when the weather is warm and humid. Spray with myclobutanil at green cluster, pink or white bud, petal fall and 14 days after petal fall. Rake up and burn fallen leaves in autumn – prune scabby shoots in winter.

## Brown rot

Most tree fruit can be affected, but the disease is worst on apples. The infected fruit turns brown and concentric rings of yellowish mould appear. The shrunken mummified fruit may remain on the tree throughout the winter or it may fall. This fungal disease cannot be controlled by spraying and it is necessary to destroy affected fruit as soon as they are seen on the tree or on the ground. Remember to store only sound fruit and examine them at regular intervals.

## Russetting

The rough scurf which sometimes forms over the surface of apples and pears is known as russetting. On some apple varieties it is natural; on the smooth-skinned varieties it is a disorder which make the fruit unsightly. The eating quality is not affected. Poor growing weather is the most likely cause, especially if it occurs at petal fall. Drought, mildew, starvation and too little organic matter in the soil are other possible causes of russetting.

## Bitter pit

Small brown areas appear on the surface tissue, each one marked by a small depression in the skin. The brown areas are occasionally scattered throughout the fruit, rendering it bitter and inedible. Bitter pit usually develops during storage, although with *Bramley's Seedling* symptoms often show while the apples are still on the tree. The cause of this disorder is complex, but it is linked to calcium deficiency and a period of water shortage. Mulching and watering in summer is helpful, but avoid using straw or strawy manure. Do not drastically prune trees which have been affected by bitter pit. If a regular problem try spraying with calcium nitrate at fortnightly intervals from mid-June.

## Codling moth

The pale pink grub bores into developing fruit and feeds on the central core. Pears and plums may be attacked as well as apples. Grubs can be found inside the fruit in July and August; the tell-tale sign is sawdust-like 'frass' within the apple. Spray with pyrethrins or plant oils/extracts in mid-June; repeat 3 weeks later.

## Apple sawfly

A ribbon-like scar is produced on the surface. Later the creamy-white grub burrows down to feed on the central core which generally causes the fruit to drop in June or July. The grubs go into the ground in July. Sticky 'frass' can be seen around the hole. Spray with pyrethrins or plant oils/extracts at petal fall. Pick up and burn the fallen apples.

## Pocket plum

A fungal disease, visible by midsummer, which prevents development of the stone and causes the fruit of plums and damsons to develop abnormally. Later, a white covering of fungus appears on the fruits which then shrivel and fall. Chemical control is not usually necessary but spraying with a copper-based fungicide after flowering may help.

## Capsid bug

Young fruitlets of apples and pears show reddish-brown spots following attack by these active green bugs. As the fruits enlarge, pale corky patches develop and the surface may be distorted, cracked and russetted. Despite the unsightly appearance the flesh remains sound. Leaf damage by capsid bugs is prevented by spraying at green cluster (page 52). To protect fruit, spray with thiacloprid, lambda-cyhalothrin or deltamethrin at petal fall.

## Fruit drop

Fruitlets will fall following insect damage – look for grubs in fallen apples, pears and plums. But healthy fruitlets also fall – this may actually be beneficial as a heavy set must be thinned. The first drop of apples takes place when the fruitlets are pea-sized – the usual reason is incomplete pollination due to a cold wet spring. The major shedding is the 'June drop'. There are several reasons why the fall at this time may be unusually heavy with only a spare crop remaining. It is normal for newly planted trees to shed most of their fruit in this way, and some varieties, such as *Cox's Orange Pippin*, have a notoriously heavy June drop. Irregular water supplies at the root, starvation, frost damage and overcrowding are other causes.

# EARLY WARNING

Many damaging pests can be controlled if they are dealt with early, before they can multiply and become a serious problem. However, knowing when a pest is likely to be present and when to get the best control can be difficult to gauge. Traps can be used to monitor the activity of some insect pests and can be useful as a warning that the pests are present. They indicate that any measures, such as sprays, applied at this time will give the best control.

## Pheromone traps

These contain a lure based on the odour of unmated females, which attracts male insects into a container where they are held until they can be counted. Increasing numbers act as an early-warning system to let you know when to control the pest. Similar traps have also proved very useful for monitoring the presence of newly hatched larvae, allowing sprays to be applied at the most effective time. Codling moth and plum moth can be monitored in this way.

## Grease bands

Traps like grease bands act as a physical barrier and can be used to prevent crawling pests, such as the female winter moth, reaching their preferred breeding site higher in the tree. Using this method of control will greatly reduce the population of pests.

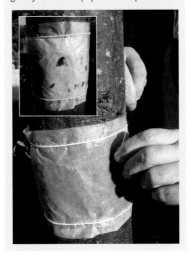

## Other traps

Other traps, like the 'Kairomone' trap, work in a slightly different way, by giving off the odour of the host plant to attract and trap both males and females. Codling moth can be controlled using this method.

These scent-based traps are far less effective as a control than the grease band, but are very good indicators that other treatments will be effective. The positive side of such a trap is that, because the lure is designed to attract a particular species of insect pest, there is less chance of beneficial insects being affected by mistake.

# Chapter 3

# SOFT FRUIT

'Soft fruit' is a general term for a widely diverse group of plants which are generally head-high or smaller and which bear soft-skinned juicy fruit – the classification of the group is shown on page 4. Not all types fit into this definition – grapes and kiwi fruit can be tall and melons have a thick rind.

The increase in shop prices and the interest in home freezing have combined to make soft-fruit growing much more popular than ever before. Unlike vegetables, you can crop plants year after year once they are established, and unlike tree fruit there is a place for at least one example in every garden. Plant gooseberry bushes and raspberry canes where space permits, or plant strawberries in a tub or window box where the garden is tiny. All are self-fertile, pruning is easier than with tree fruit and new plants quickly start to bear fruit – within a year for melons and some strawberries.

The attraction of luscious soft fruit from one's own garden is obvious, but beginners do run into problems because they ignore two basic points:

● **Soft fruit needs attention** You cannot neglect the plants during the year and still expect long-living and high-yielding bushes and canes. This does not mean that soft fruit care is difficult, but you must pay attention to feeding, pruning and spraying. The plants are prey to all sorts of pests and problems – keep watch and take remedial action at the first sign of trouble. Two problems overshadow all the others – birds and viruses.

● **There are not many general rules** A small number of rules apply to all types. Enrich the soil with well-rotted manure or compost before planting and feed each year in spring. Keep the bushes or canes free from weeds and make sure there is ample water around the roots after planting and when the fruits are swelling. But the differences in looking after the various types of soft fruit far outweigh the similarities. Study the planting and pruning instructions for each type – even blackcurrant and red currant are treated quite differently.

So soft fruit is for everyone. The usual choice is strawberries and raspberries – you will find popular ones like *Cambridge Favourite* and *Glen Clova* everywhere. But do try to be a little more adventurous – there are exciting new types and varieties in the catalogues. Replacement will be necessary after a time – strawberries should remain productive for 3–4 years and raspberries for 10–12 years. You can propagate your own replacement stock, but do make sure the plant is virus-free before taking cuttings or runners. This is not as easy as it sounds, as plants may be infected and yet show no obvious symptoms.

## Buying

The rules for buying tree fruit are laid down on page 8. Some of the basics apply here – avoid 'bargains' at all costs and go to a reputable supplier. Order early before the best varieties are sold out, and choose container-grown canes or bushes rather than bare-rooted ones if you have missed the autumn–spring planting period.

The need for care when buying is greater with soft fruit than with tree fruit. Virus-infected stock will never amount to anything, so do not plant gifts from friends and always go for certified stock where the scheme applies.

### Certified Stock

There is a certification scheme for strawberries, raspberries, blackberries and hybrid berries, gooseberries, blackcurrants, red and white currants. Certified plants have been inspected and passed by Department for Environment, Food and Rural Affairs officials as substantially free from pests and diseases, and also as true to type. Unhealthy-looking plants are rejected.

## THE THINGS TO LOOK FOR

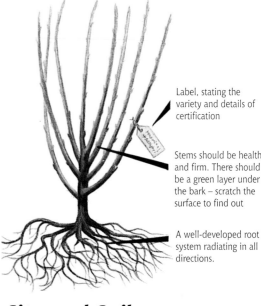

Label, stating the variety and details of certification

Stems should be healthy and firm. There should be a green layer under the bark – scratch the surface to find out

A well-developed root system radiating in all directions.

## Site and Soil

Ideally soft fruit should be grown in full sun, but nearly all types will do reasonably well in light shade. Most localities and soils are suitable, but there are some sites which will give disappointing results. These include land exposed to strong winds or late frost; gravelly, sandy or poorly drained soil; and ground under trees. Avoid planting near fruit trees which may need spraying when soft fruit is being harvested and plant tall canes and bushes away from low-growing crops where shade could be a problem. Walls and fences can be used for bush fruit grown as fans or cordons where space is short. Strawberries can be grown in containers.

# Planting

Follow the general rules set out for tree fruit on pages 9 and 10. Make sure the ground is prepared thoroughly – enrich with organic matter and remove roots of perennial weeds. Check on any special needs for the types of soft fruit to be planted – some don't like acid conditions but others will fail miserably if the land is not acidic.

Never replace an old type of soft fruit with a bush or cane of the same type – for example, do not plant raspberry canes where you have just removed 10-year-old canes of the same or a different variety of raspberry. Find a spot where soft fruit has not been grown before if you can, or else grow a different type of soft fruit when replacing old plants.

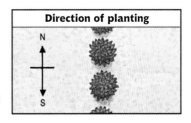

**Direction of planting**

N

S

# The Bird Problem

Birds are a big problem. Fruit buds are attacked in winter and spring – the flowers which survive produce fruits which when ripe are again attacked. Some people plant a few extra bushes for the birds to try to get round the problem, but there are times when you will get little or no fruit if you don't do something about the bird menace.

Raspberries and strawberries are the types most at risk. Chemical sprays are of little use and mechanical bird scarers are of limited value. The only satisfactory solution is to erect a fruit cage. For strawberries and low-growing bushes and canes a temporary cage will do, but for standard-size bushes and canes a walk-in cage is a better idea.

## Temporary cage

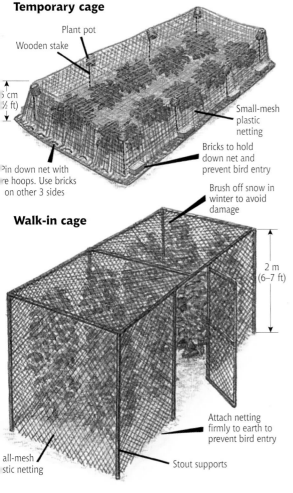

Plant pot

Wooden stake

5 cm
½ ft)

Pin down net with
re hoops. Use bricks
on other 3 sides

Small-mesh
plastic
netting

Bricks to hold
down net and
prevent bird entry

Brush off snow in
winter to avoid
damage

## Walk-in cage

2 m
(6–7 ft)

Attach netting
firmly to earth to
prevent bird entry

all-mesh
stic netting

Stout supports

# The Virus Problem

Viruses are not just another soft-fruit disease – they are the disease which threatens the productive life of canes and bushes. Viruses are carried from plant to plant by insects such as greenfly and leafhopper and once infected the plant cannot be cured. The symptoms depend on both virus and plant type – leaves may be blotched, mottled or streaked and growth may be distorted or dwarfed. In some cases the infection may prove fatal. The usual response is for the health and fruitfulness to decline so that removal and burning of the infected plant is the only thing to do.

Tackle the virus problem by using certified stock and then spraying to keep the insect carriers under control. In this way productive cropping should continue for at least the 10-year life span we can expect from bush and cane soft fruit.

# Feeding and Mulching

A spring dressing of fertilizer has 2 functions – it promotes a good crop later in the season and also encourages the production of new growth which will bear next year's crop. Treat an area extending about 1 m (3 ft) from the stems – use a general fertilizer such as Growmore or pelleted chicken manure at 60 g per sq. metre (2 oz per sq. yard) or apply a high-potash liquid fertilizer at the manufacturer's recommended rate. Most fruit responds to foliar feeding, especially after pest damage and when the small fruits are beginning to swell. Yellowing leaves in alkaline soil are a symptom of iron or manganese deficiency. Spray the plants with seaweed extract.

The purpose of mulching is to keep down weeds and to conserve moisture. Black polythene sheeting can be used – it doesn't break down and it fully suppresses weeds. Organic mulches do not share these virtues, but they do provide nutrients and help to improve soil structure. Lay down a layer of shredded bark, well-rotted manure or compost – this should be about 5–8 cm (2–3 in.) thick and it should extend for about 1 metre (1 yard) around the stems. Keep the ground clear for a few centimetres/inches around the stems – place mulches in position immediately after weeding and feeding in spring.

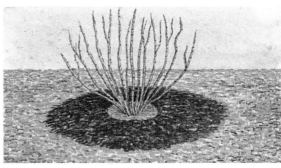

# STRAWBERRIES

Strawberries are the most popular soft fruit for growing at home. This should come as no surprise, as there are so many features in their favour. You don't need lots of room – you can grow them in the vegetable plot or in a container on the patio. You also don't need lots of patience – plant in August and you can be picking fruit in June or July next year. And the result is luscious and attractive fruit, each one weighing up to 55 g (2 oz) and each plant yielding 170–700 g (6–24 oz) of strawberries.

The problem is that the season is all too fleeting. If just one popular variety is grown, picking in June or July will last for two or three weeks and then it is over. But if you have the space and choose the varieties with care, it is possible to pick strawberries for six months of the year – from May until the arrival of the first frosts. By covering the plants with cloches in February you can pick about three weeks earlier – by growing in a heated greenhouse you can have fruits at Christmas.

The summer-fruiting varieties are the most popular and dominate the catalogues. Just one large flush is produced – the early and 2nd early types are in fruit between late May and late June; the mid-season ones range from mid-June to mid-July; and the late types are ready between early and late July. The first step is to buy young, healthy plants in late summer or early autumn – plant no later than mid-September to ensure that a good root system can develop before the onset of winter.

Perpetual varieties extend the season. Several flushes are produced, the main ones being between August and October. You never get much at a single picking, but the out-of-season crop is very welcome. Recently a new group appeared – the day-neutral varieties. Here fruiting is not controlled by day length and there is a simple message – pick 12 weeks after planting. In practice, however, they behave like perpetual varieties and the flavour is unfortunately indifferent. Finally, there are the alpine strawberries, tiny berries which appear in modest amounts over a prolonged period – you will need 20–30 plants for a worthwhile crop.

Look for something different in the catalogues if you are adventurous. There are pink-flowered types and others with pink or yellow fruit. For most of us, however, it is better to stick to old favourites and highly recommended new ones.

### Summer-fruiting strawberry

(Other names: **Single-crop, June-bearer**)
By far the most popular group. Largest and best-quality fruits found here.
Cropping time between late May and late July. Single flush – a few varieties occasionally produce a second crop.
Productive life: 3–4 years

### Alpine strawberry

(Other name: **Wild**)
Fruit is very small – aromatic and sweet but not juicy.
Grow in a container or as an edging plant.
Sow seed in autumn – plant out in spring. Cropping time midsummer to late autumn. Red, yellow and white varieties are available.
Productive life: 1 year

### Perpetual strawberry

(Other names: **Remontant, Everbearer, Autumn-fruiting, Two-crop**)
Fruit is generally smaller and less sweet than summer-fruiting varieties – plants are less hardy. Cropping time June and again in late summer–autumn. Remove early flowers to increase the autumn crop.
Productive life: 2 years

### Day-neutral strawberry

Only a few varieties available.
Outdoors they behave like ordinary perpetuals in Britain – under glass they will produce winter fruit 12 weeks after planting.
Remove early flowers in the garden to increase fruit size and yield of the autumn crop. Productive life: 2 years

# Varieties A–Z

### Alexandria

**Type:** Alpine

**Fruit size:** Small

**Skin colour:** Bright red

**Picking time:** Midsummer–late autumn

This variety bears the largest fruits amongst the alpine strawberries – about 1 cm (½ in.) long. The colourful fruits are also juicier than other alpines, but the yields are less than you will get from the more popular *Baron Solemacher*. The bushy plants do not produce runners. Flavour is excellent.

### Anablanca

**Type:** Alpine

**Fruit size:** Small

**Skin colour:** White-tinged pink

**Picking time:** Early June–mid-July

This very old French variety, related to the *Pineberry*, produces white berries with an aromatic, pineapple flavour. An early-cropping variety, these small plants (by modern strawberry standards) are very hardy with foliage that is evergreen in mild climates.

**Anablanca**

### Aromel

**Type:** Perpetual

**Fruit size:** Medium–large

**Skin colour:** Medium red

**Picking time:** Late summer–autumn

The catalogues are quite right when they say that the flavour of this variety is outstanding. It is becoming popular, but there are a few drawbacks. Crops are moderate, the leaves are susceptible to mildew and the rather soft, conical fruits are often misshapen. Water well in dry weather.

### Baron Solemacher

**Type:** Alpine

**Fruit size:** Very small

**Skin colour:** Dark red

**Picking time:** Midsummer–late autumn

The most popular alpine strawberry – this variety actually prefers partial shade to full sun. Each fruit is tiny, but they are borne in large numbers over a long period. Birds seem to leave them alone. The small bushes are erect and do not bear runners. Fine flavour.

**Aromel**

### Buddy

**Type:** Perpetual

**Fruit size:** Large

**Skin colour:** Dark red

**Picking time:** July–late September

A superb variety for the home gardener, with traditional-shaped berries with a small, light green calyx that is easily removed. It shows good disease resistance and has a long cropping season.

### Cambridge Favourite

**Type:** Summer fruiting

**Fruit size:** Medium

**Skin colour:** Orange-red

**Picking time:** Mid-June–mid-July

The number-one choice for many years, and still a firm favourite – it crops well and is thoroughly reliable. When virus-free it has a long productive life and there is good all-round disease resistance. Consider other varieties – the whitish flesh lacks full flavour.

### Cambridge Vigour

**Type:** Summer fruiting

**Fruit size:** Large

**Skin colour:** Bright red

**Picking time:** Early June–late June

An old favourite, which still has its fans, particularly in colder areas. Fruits are conical, firm-fleshed and full of flavour. The vigorous and upright plants produce heavy crops, but mildew can be a problem. The quality quickly deteriorates, so planting every 2 years is recommended.

**Baron Solemacher**

### Chelsea Pensioner

**Type:** Summer fruiting

**Fruit size:** Large

**Skin colour:** Bright red

**Picking time:** Mid-July–mid-August

This is a late-season variety, with cropping peak concentrated late in the season. It produces heavy yields of large, bright red fruit with a very sweet flavour. Resistance to pests and diseases, even when mildew has affected other varieties.

**Cambridge Favourite**

Christine

Diamante

Elsanta

Grandee

### Christine

A new, early-cropping variety. The berries are large and sweet with excellent flavour. A vigorous grower which needs to be kept well watered. Good pest and disease resistance, but may need frost protection due to early cropping. Often produces a second crop.

**Type:** Summer fruiting
**Fruit size:** Very large
**Skin colour:** Orange-red
**Picking time:** June–early July

### Diamante

A new compact variety from California that is ideal for container and patio growing. It produces large, good-quality fruits with a sweet, juicy flavour at any time of the year if the temperature is kept at 12–13°C (55°F) or higher.

**Type:** Day neutral
**Fruit size:** Large
**Skin colour:** Pale red
**Picking time:** Early summer–autumn

### Elsanta

This is still the most widely grown variety in the UK. It is very popular with professional growers because it has the properties they require – high yields, attractive glossy fruit and excellent shelf life. The flavour is outstanding but disease resistance is not.

**Type:** Summer fruiting
**Fruit size:** Large
**Skin colour:** Orange-red
**Picking time:** Mid-June–mid-July

### Elvira

A recent introduction that is ideal for growing under protection if you want an earlier crop. Yields are heavy, the large conical fruit has an excellent flavour. The growth habit is open but mildew can still be a problem. One of the best varieties for forcing or growing in pots.

**Type:** Summer fruiting
**Fruit size:** Large
**Skin colour:** Dark red
**Picking time:** Early June–late June

### Flamenco

Produces heavy yields of top-quality fruit, with peak cropping in September. The large, attractive berries have a sweet flavour and juicy texture. Plants show good pest and disease resistance and perform well when grown under protection or in containers.

**Type:** Perpetual
**Fruit size:** Large
**Skin colour:** Orange-red
**Picking time:** July–October

### Grandee

Enormous fruit when well grown – specimens weighing 85 g (3 oz) have been picked. Yields are also very high, especially in the second year. One for the show bench, of course, but fruit lacks flavour. Not in many catalogues and often listed under 'Heritage' varieties these days.

**Type:** Summer fruiting
**Fruit size:** Very large
**Skin colour:** Dark red
**Picking time:** Early June–late June

### Hapil

This Belgian variety is favoured by professional growers as a replacement for *Cambridge Favourite* – with an upright growing habit, it does much better than its rival in a dry season. Heavy yields of large, glossy, conical fruit, with a sweet flavour. Well worth growing on lighter soils.

**Type:** Summer fruiting
**Fruit size:** Very large
**Skin colour:** Orange-red
**Picking time:** Mid-June–mid-July

### Honeoye

An important early variety with an awful name, but a good performance. It produces heavy crops for a summer-fruiting variety and the glossy fruit is very attractive. The firm, red flesh has a better flavour than *Pantagruella* and it has better resistance to botrytis.

**Type:** Summer fruiting
**Fruit size:** Medium
**Skin colour:** Bright red
**Picking time:** Early June–late June

### Malling Centenary

**Type:** Summer fruiting

**Fruit size:** Medium–large

**Skin colour:** Orange-red

**Picking time:** Early June–early July

A new British-bred heavy cropper. The flavour is good and the attractive fruits very juicy and sweet with a round or conical shape. The fruits hang away from the foliage, making picking much easier. This variety has no apparent susceptibility to pests and diseases.

### Malling Opal

**Type:** Perpetual

**Fruit size:** Large

**Skin colour:** Mid-red

**Picking time:** Early August–October

The fruit has a conical shape and attractive appearance with a juicy texture and sweet flavour. It can carry on cropping until Christmas when grown under protection. It has some resistance to powdery mildew and is very much an ideal home-garden variety.

### Malwina

**Type:** Summer fruiting

**Fruit size:** Large–very large

**Skin colour:** Red–dark red

**Picking time:** Mid-July–mid-August

This is becoming a popular late-cropping variety, easy to grow with good overall disease resistance. It gives good yields, is dependable and easy to pick. The large red fruits have a good flavour and exceptional sweetness.

### Pandora

**Type:** Summer fruiting

**Fruit size:** Large

**Skin colour:** Orange-red

**Picking time:** Mid-July–mid-August

A late-cropping, high-yielding variety with large fruits of excellent flavour, and a pleasant juicy texture. The vigorous plants appear to have good disease resistance, but it is unique in requiring a pollination partner such as *Cambridge Favourite*.

### Pantagruella

**Type:** Summer fruiting

**Fruit size:** Medium–large

**Skin colour:** Orange-red

**Picking time:** Late May–mid-June

This is the one to grow to beat your neighbours – *Pantagruella* is the first strawberry to bear fruit. Place cloches over the plants and you can enjoy the fruit in early May. Nothing special apart from earliness – average flavour, disease-prone, frost-prone and poor after the first year.

### Pegasus

**Type:** Summer fruiting

**Fruit size:** Large

**Skin colour:** Orange-red

**Picking time:** Late June–late July

A new variety from East Malling Research Station in Kent. It has excellent disease resistance. This is a heavy cropper producing very good-quality dessert fruit with tender flesh and a sweet, juicy flavour. The fruits are a good shape with a glossy finish to the skin.

### Rapella

**Type:** Perpetual

**Fruit size:** Medium–large

**Skin colour:** Bright red

**Picking time:** Late summer–autumn

A newer variety tipped to challenge *Aromel* for the perpetual crown. It crops more heavily than others in the group and the flavour is good. Fruits ripen earlier than other perpetuals, so all can be picked before the first frosts. Susceptible to mildew – not suitable for northern areas.

### Red Gauntlet

**Type:** Summer fruiting

**Fruit size:** Large

**Skin colour:** Bright red

**Picking time:** Mid-June–mid-July

Well established – it's been around for many years. Cropping is good but the flavour is disappointing. The disease resistance is quite high, but its main claim to fame is the second crop which appears in September after a hot summer. Still popular, but not the best choice.

Pantagruella

Pegasus

Rapella

Red Gauntlet

**Royal Sovereign**

**Selva**

**Tamella**

**Tenira**

### Royal Sovereign

This is the only pre-war variety in the catalogues, and that's mainly for sentimental reasons. Once considered to be the best-flavoured fruit, it is equalled nowadays by *Elsanta*. Does well on the heavier types of soil, but it is not a heavy cropper. Disease resistance is poor.

**Type:** Summer fruiting
**Fruit size:** Medium–large
**Skin colour:** Orange–red
**Picking time:** Early June–late June

### Saladin

This Scottish variety has produced very heavy yields in trials. The fruit is often ribbed and misshapen and the surface is soft. Growth is erect and leafy, and it is resistant to red core and mildew. The flavour is 'poor', 'average' or 'very good', depending on the catalogue.

**Type:** Summer fruiting
**Fruit size:** Large
**Skin colour:** Orange–red
**Picking time:** Early July–late July

### Selva

One of the day-neutral varieties which will crop several months after planting at any time of the year if the temperature is kept at 12–13°C (55°F) or higher. The fruit is unusually firm and attractive but the flavour is disappointing. Susceptible to mildew.

**Type:** Day neutral
**Fruit size:** Very large
**Skin colour:** Bright red
**Picking time:** Late summer–autumn

### Sweetheart

A new variety, which provides heavy crops of large, conical fruits. The flavour is pleasant, with a balance of sweetness and acidity. The plants have a compact habit with the fruit well displayed. Mildew-resistant foliage. One suitable for container growing.

**Type:** Summer fruiting
**Fruit size:** Large
**Skin colour:** Light red
**Picking time:** Mid-June–mid-July

### Tamella

A good choice, especially for northern England and Scotland. It crops over a long period and the productive life of the plant is longer than for nearly all other varieties. It also gives the highest yields, but the flavour of the orange flesh is only fair.

**Type:** Summer fruiting
**Fruit size:** Very large
**Skin colour:** Dark red
**Picking time:** Mid-June–mid-July

### Tenira

Grow this one if you need to be convinced that a modern variety can taste as good as a pre-war one. Plants are vigorous and disease resistance is better than average, but yields are only moderate. Growth is upright, so ripe fruit is kept off the ground.

**Type:** Summer fruiting
**Fruit size:** Medium
**Skin colour:** Bright red
**Picking time:** Early July–late July

### Totem

Yields are much lower than you would get from *Cambridge Favourite* and the flavour when eaten fresh does not compare with *Elsanta* or *Hapil*. The sole reason for choosing this mid-season variety would be for its outstanding freezing properties – it keeps its shape when thawed.

**Type:** Summer fruiting
**Fruit size:** Medium
**Skin colour:** Dark red
**Picking time:** Mid-June–mid-July

### Troubador

A useful variety to extend the season – cropping starts about a fortnight after *Cambridge Favourite*. Yields are high and disease resistance is very good, but flavour is only average and the plants suffer in a dry season. The glossy fruit is attractive. Late flowering – an advantage in northern areas.

**Type:** Summer fruiting
**Fruit size:** Medium–large
**Skin colour:** Dark red
**Picking time:** Early July–late July

# Planting

A sunny, sheltered site is required – a little shade during the day is acceptable. The area chosen for the strawberry bed must not be in a frost pocket and ideally the land should not have been used for growing potatoes, tomatoes, chrysanthemums or strawberries for several years. Good drainage is essential and light soil will ensure early crops.

The ground should be slightly acid and a liberal amount of organic matter is required.

Dig at least a month before planting – all weeds must be removed at this stage. Mix in a bucketful of garden compost or well-rotted manure (10 litres per sq. metre/2 gallons per sq. yard) when preparing the bed. Scatter a general fertilizer such as Growmore at 60 g per sq. metre (2 oz per sq. yard) over the ground just before planting.

## Summer-fruiting varieties

Plant bare-rooted plants between July and September. March–April planting is acceptable, but all flowers should be removed during the first season.

Note: Container-grown specimens can be planted at any time of the year when the weather is suitable.

Space between plants: 45 cm (1½ ft)

Space between rows: 75 cm (2½ ft)

## Perpetual and day-neutral varieties

Plant in July–August or March–April. Remove the first flush of flowers during the first season.

Note: Container-grown specimens can be planted at any time of the year when the weather is suitable.

Space between plants: 45 cm (1½ ft)

Space between rows: 75 cm (2½ ft)

## Alpine varieties

Sow seed in autumn – overwinter in an unheated greenhouse. Plant out home-grown or bought-in plants in April or May.

Space between plants: 30 cm (1 ft)

Space between rows: 60 cm (2 ft)

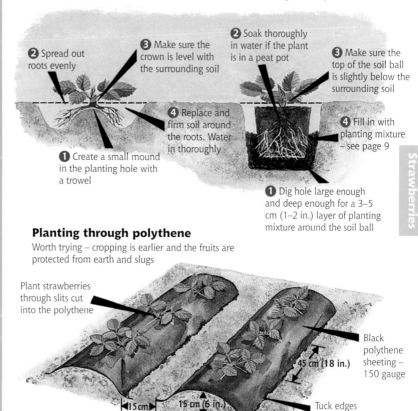

**Bare-rooted plants**

❷ Spread out roots evenly

❸ Make sure the crown is level with the surrounding soil

❹ Replace and firm soil around the roots. Water in thoroughly

❶ Create a small mound in the planting hole with a trowel

**Container-grown plants**

❷ Soak thoroughly in water if the plant is in a peat pot

❸ Make sure the top of the soil ball is slightly below the surrounding soil

❹ Fill in with planting mixture – see page 9

❶ Dig hole large enough and deep enough for a 3–5 cm (1–2 in.) layer of planting mixture around the soil ball

**Planting through polythene**

Worth trying – cropping is earlier and the fruits are protected from earth and slugs

Plant strawberries through slits cut into the polythene

Black polythene sheeting – 150 gauge

45 cm (18 in.)

15 cm (6 in.)

15 cm (6 in.)

Tuck edges into the soil

# Seasonal Care

For a May crop grow an early variety such as *Pantagruella*, *Honeoye*, *Cambridge Vigour* or *Malling Centenary* and cover with cloches from late winter onwards. When the flowers appear, partially open the cloches during the day to enable pollinating insects to enter.

Cover unprotected plants with newspaper if frost is expected when the flowers are open. Remove the paper during the day.

Newly planted strawberries must be watered regularly – so must established plants during dry weather. It is important to keep water off fruits – do this task in the morning so splashes can dry before nightfall.

Strawberries are low-growing plants – yields can be seriously affected by abundant weed growth. Keep weeds under control by hoeing regularly – do not hoe too deeply and keep away from the crowns.

May is the usual time for mulching – see 'Feeding and Mulching' overleaf. Make sure that weeds are removed and the soil is moist before putting down a mulch. After mulching it will be necessary to provide bird protection – see page 63.

Most varieties produce runners which bear plantlets along their length. These runners can be removed so that each plant keeps its individuality, or you can let the runners grow and root between the plants to form a continuous ('matted') row. Mulching and weeding can be more difficult here and the fruits tend to be smaller than on runner-removed plants, but the total yields are much higher.

Spray as necessary to keep down pests and diseases. After removing the leaves in summer or autumn (see 'Pruning' overleaf), hoe to remove weeds and excess runners and to loosen the soil between the plants.

# Pruning

Pruning takes place neither at planting time nor when the plants are growing. The time for pruning is when the last of the berries have been picked. With summer-fruiting varieties cut off all the leaves about 7–8 cm (3 in.) above the crown. At the same time cut off unwanted runners and rake up together with mulching material, weeds, etc., and dispose of by composting or burning. Less drastic pruning is required with perpetual and day-neutral varieties – remove old leaves only.

# Feeding and Mulching

Apply a general fertilizer such as Growmore at 60 g per sq. metre (2 oz per sq. yard) around the plants in March or April – keep granules off the foliage. After fruiting apply a high-potash feed.

Mulching is essential to prevent the ripening fruit from coming in contact with the soil. The traditional mulching material is barley straw, which is placed along the rows and tucked under the fruit. Alternatively, use pieces of black polythene sheeting or proprietary strawberry mats. Mulch when the fruit is beginning to swell – place slug pellets under the plants before putting the mulch in place.

# Picking and Storage

Once maximum size is reached the fruit ripens very rapidly. You must inspect the bed daily and pick strawberries that have reddened all over. Do this in the morning and when the fruit is dry. Nip the stalk between thumbnail and forefinger – don't tug the fruit away from the stem. Remove and destroy diseased fruit immediately.

Eat as soon as possible – do not remove plugs (cores) before washing or storing. Strawberries can be kept for a few days in a cool place, but they freeze badly. If you want to freeze them for serving whole rather than as a purée you should choose the variety *Totem* or use small, slightly unripe fruit.

# Growing under Glass

Strawberries grown in pots in an unheated house will fruit in late April or early May – choose *Elvira* or *Cambridge Favourite*. Put vigorous, well-rooted plantlets into 13–15 cm (5 or 6 in.) pots in August. Keep them in the open – water regularly and then bury the pots in well-drained ground in November. Transfer the containers to a cold greenhouse in February – hand pollinate when the flowers appear and start to feed weekly with a liquid fertilizer. Ventilate the house when the weather is warm and sunny. Stop feeding when fruits start to redden.

# Raising New Plants

New strawberries can be easily raised from runners, but you must make sure that the mother plant is healthy, with leaves which are neither crinkled nor mottled. In June or July select 4 or 5 strong runners from each plant and proceed as shown below:

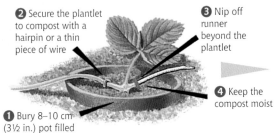

❷ Secure the plantlet to compost with a hairpin or a thin piece of wire

❸ Nip off runner beyond the plantlet

❹ Keep the compost moist

❶ Bury 8–10 cm (3½ in.) pot filled with compost

❺ After 4–6 weeks the plantlet will have rooted. Sever plantlet from runner and remove hairpin

❻ About 7 days later remove rooted plantlet from the pot – plant as instructed on page 69

Some perpetual and day-neutral varieties do not produce runners. Propagate by division – dig up the plant in early September and divide into a number of separate crowns. Plant each one immediately – the crown of each divided plant should be level with the soil surface after planting.

# Strawberries in Containers

A problem with many fruit trees, canes and bushes is the need for space and/or free-draining soil. This is hard for the person with a tiny plot or no garden at all – but strawberries pose no difficulty. They can be grown in tubs, window boxes, hanging baskets, growing bags, etc.

The traditional strawberry pot is a large terracotta jar with cupped holes arranged spirally around its sides. An attractive feature in a cottage garden or some other old-world setting, but perhaps a little out of place on the terrace of a modern house. More fitting here is the plastic tower or the polished wooden tub. The DIY enthusiast can make a strawberry barrel – drill a series of 5 cm (2 in.) diameter holes about 30 cm (1 ft) apart.

There are cultural as well as decorative and space-saving benefits. The fruit is kept away from slugs and picking is made easier. Mulching and hoeing are not required, and bird protection simply involves draping a net over the container. There is an extra task, however – you will have to water regularly.

# RASPBERRIES

Raspberries continue to play second fiddle to strawberries in the gardens of Britain. There are some obvious reasons – the plants are much taller, which makes bird protection more difficult. Strong supports are necessary, and so is annual pruning. Still, there's the other side to the coin. Raspberries freeze extremely well, and canes can be expected to remain productive for 8 to 12 years. The cropping period for summer-fruiting varieties lasts from 3 to 6 weeks – almost double the strawberry season. An even more important advantage is the late flowering of raspberries – frost damage is rarely a problem. No worries here for the northern gardener – raspberries thrive in cool, damp summers.

The summer-fruiting varieties are by far the most popular type. The root ('stool') of each plant produces tall canes – fruit is borne on those canes which developed in the previous year. Picking early varieties begins in early July – mid-season varieties start to crop in mid- or late July and late varieties are not ready until late July or early August.

To make a start you will need land which does not waterlog in winter – if drainage is poor, create a raised bed. This site should not have grown raspberries for 7 or more years – if it has then change the soil to a spade's depth. You will need some form of support – a stout fence, a line of strong poles or a custom-made post-and-wire structure. Finally, the plants – these must be bought from a reputable supplier and be certified as virus-free.

Autumn-fruiting varieties can be used to extend the cropping season from late August to the arrival of the first frosts. These raspberries fruit on the top of this year's canes, which means that they are pruned differently from the more popular summer-fruiting types. Autumn raspberries have smaller canes and smaller yields – they do best in mild regions.

Raspberries are an easy crop and very rewarding – only strawberries start to bear fruit more quickly and give a higher yield per area of bed. Birds and viruses are the major menaces. Deterioration usually starts after about 8 years – matters get steadily worse and after a few more years (or even less) it will be necessary to dig them out and burn the virus-ridden plants.

Consider the newer varieties when replanting – look for aphid resistance. Grow *Malling Admiral* if you have room for only one variety. *Autumn Bliss* should be your first choice amongst autumn-fruiting varieties – it's so much better than some of the older ones.

## Summer-fruiting raspberry

Yellow    Orange-red    Pink    Red    Dark red

● By far the more popular group. Yields are higher and plants can be grown in all districts.

● Fruit is borne over a 3–6 week period in July and/or August, depending on the variety.

● Fruit is borne on canes produced in the previous year. Summer-fruiting varieties are usually referred to as floricane, because they have stems that grow for one year before producing lateral shoots bearing flowers and fruit the following year.

● Support is essential.

● Protection against birds is essential.

● Removal of fruit from the plug is very easy at picking time.

● Average yield: 2–2.25 kg per metre of row (4½ lb per yard of row).

## Autumn-fruiting raspberry

Yellow      Red      Dark red

● Less popular than summer-fruiting varieties. Yields are lower and plants do not thrive in cold districts.

● Fruit is borne over an 8–9 week period between late August and the first frosts, depending on the variety. You may see autumn-fruiting varieties referred to as primocanes because they produce fruit in their first year of growth.

● Fruit is borne on the tips of canes produced this year.

● Support is desirable but not essential.

● Protection against birds is desirable but is not always essential.

● Removal of fruit from the plug is often difficult at picking time.

● Average yield: 0.7–0.75 kg per metre of row (1½ lb per yard of row).

**Autumn Bliss**

**Delight**

**Fallgold**

**Glen Clova**

# Varieties A–Z

### Autumn Bliss
A newer autumn-fruiting raspberry which has become the recommended choice. It significantly out-yields many of the other varieties and the fruits start to ripen in August. The texture is firm and the flavour is quite good. Sturdy – support is not needed. The site must be sheltered and sunny.

**Type:** Autumn fruiting
**Fruit size:** Medium–large
**Skin colour:** Bright red
**Picking time:** Mid-August–October

### Erika
A newer, high-yielding variety from Italy, which has the potential to crop twice a year as it can produce fruit on both the current and previous year's canes. Canes are vigorous, tall and erect, reaching up to 2 m (6 ft). The fruits have a sweet, juicy flavour and are clearly visible, making picking easier.

**Type:** Autumn fruiting
**Fruit size:** Large
**Skin colour:** Orange-red
**Picking time:** September–October

### Fallgold
This is a yellow raspberry which fruits in autumn. A novelty, but the unique colour is not matched by any other unique properties. The flavour and sweetness are good without being outstanding – so are fruit size and yields. Not generally available at garden centres.

**Type:** Autumn fruiting
**Fruit size:** Medium–large
**Skin colour:** Yellow
**Picking time:** Mid-September–Octobe

### Glen Clova
A very popular variety – yields are high and the fruits herald the start of the raspberry-picking season in many gardens. The berries are firm and rather small, making them ideal for freezing. Not perfect – flavour is only fair and the plant is susceptible to virus. Plant it away from other varieties.

**Type:** Summer fruiting
**Fruit size:** Small–medium
**Skin colour:** Red
**Picking time:** Early July–late July

### Glen Moy
A newer variety to compete with *Glen Clova*. There are advantages – no spines, larger berries, good aphid resistance, a small autumn crop in warm summers and a better flavour. Gives good yields, but the picking season is quite short. The fruit is conical with a downy skin.

**Type:** Summer fruiting
**Fruit size:** Medium–large
**Skin colour:** Pale red
**Picking time:** Early July–late July

### Glen Prosen
A newer variety to compete with *Malling Admiral*. The fruit is round and firm; the canes are spine-free and easy to control. Flavour is good and there is some resistance to both aphids and virus. A heavy-cropping variety, but it needs better soil than most raspberries.

**Type:** Summer fruiting
**Fruit size:** Medium
**Skin colour:** Bright red
**Picking time:** Mid-July–mid-August

### Joan J AGM
This outstanding variety produces high-quality fruits with an excellent flavour and fleshy texture. The numerous canes are stout and spine-free, vigorous with a tall and erect habit. The earliness, large fruit size and excellent eating qualities make this variety an ideal choice for gardeners keen to grow their own.

**Type:** Autumn fruiting
**Fruit size:** Very large
**Skin colour:** Bright red
**Picking time:** Late July–October

### Joy
A summer-fruiting variety which neatly fills the cropping gap between *Glen Clova* and *Leo*. It has a number of good points – it crops for about 6 weeks, yields are high and flavour is very good. There is some aphid resistance. No real problems, but stems are very spiny.

**Type:** Summer fruiting
**Fruit size:** Large
**Skin colour:** Dark red
**Picking time:** Mid-July–late August

### Leo AGM

**Type:** Summer fruiting
**Fruit size:** Medium–large
**Skin colour:** Orange-red
**Picking time:** Late July–late August

This is the last of the popular summer-fruiting raspberries to bear fruit. The taste is good and quite tangy – texture is firm and growth is vigorous. Stems are both aphid- and botrytis-resistant. The major problem is that the canes are rather sparse and the yields are below average.

### Malling Admiral AGM

**Type:** Summer fruiting
**Fruit size:** Large
**Skin colour:** Bright red
**Picking time:** Mid-July–mid-August

Choose this one if you are restricted to one variety. The spine-free canes are numerous, vigorous and resistant to spur blight, botrytis and some viruses. Flavour is very good, and texture is firm and yields are reasonably high. Protect from strong winds.

### Malling Jewel AGM

**Type:** Summer fruiting
**Fruit size:** Medium
**Skin colour:** Dark red
**Picking time:** Early July–late July

An all-purpose and reliable variety which has been very popular for many years. The flavour is good but yields are not high. Growth is compact, making it popular for smaller gardens – it is a good idea to plant 2 canes at each station. An important virtue is tolerance to virus infection.

Malling Jewel

### Malling Orion

**Type:** Summer fruiting
**Fruit size:** Small–medium
**Skin colour:** Pink
**Picking time:** Mid-July–mid-August

An old variety which is making something of a comeback. The round fruits have a good flavour and firm texture. Growth is vigorous, yields are quite heavy, but all these properties are matched by newer varieties which have additional benefits such as aphid resistance.

Malling Orion

### Malling Promise

**Type:** Summer fruiting
**Fruit size:** Large
**Skin colour:** Orange-red
**Picking time:** Early July–late July

Shares with *Glen Clova* the honour of being the earliest raspberry in the garden. It also shares the rather insipid taste of the fruit, but yields are very good and it is tolerant of virus infection. It is an easy variety to grow and it does well in poor soil.

### Octavia

**Type:** Summer fruiting
**Fruit size:** Medium–large
**Skin colour:** Pink-red
**Picking time:** August

This variety produces a heavy yield of berries along the entire length of the upright canes. The flavour is sweet and the fleshy fruits have fewer seeds than many of the older varieties. A good choice for extending the summer-fruiting season before the autumn varieties start cropping.

Tulameen

### Tulameen AGM

**Type:** Summer fruiting
**Fruit size:** Very large
**Skin colour:** Pinkish-red
**Picking time:** Early July–late August

This Canadian-bred variety is probably the most widely grown, due to its fruit quality and high yields. The berries have a good flavour and are produced on long laterals which make for easy picking, especially as the canes are tall and spiny. The late cropping makes it ideal for northern gardens.

### Valentina

**Type:** Summer fruiting
**Fruit size:** Medium–large
**Skin colour:** Orange-pink
**Picking time:** June–July

A very striking and attractive raspberry, with apricot-pink fruit, and a sweet, juicy flavour. It produces strong, upright canes with only a few spines, making harvesting easy. With good resistance to many pests and diseases, an ideal variety for those who prefer to grow organically.

Valentina

# Planting

A sheltered spot is required – shoots can be damaged by strong winds. Raspberries grow best in full sun but partial shade is acceptable. The ideal soil is slightly acid, not too heavy and rich in organic matter. The great danger is waterlogging – raspberries rapidly die if their roots stand in wet, airless earth for long periods. Raise the proposed planting site by adding topsoil to the surface if the subsoil is solid clay.

Dig at least a month before planting – make a trench about 45 cm (18 in.) wide and 20–25 cm (8–9 in.) deep. All weeds must be removed at this stage. Add a bucketful of compost or rotted manure every metre (yard) and spread it across the bottom of the trench. Fork in and return the soil. Consolidate the surface and scatter a general fertilizer such as Growmore at 60 g per sq. metre (2 oz per sq. yard) over the ground before planting.

Plant bare-rooted canes in November or December – plant in March if this is not possible.

Note: Container-grown specimens can be planted at any time of the year when the weather is suitable.

Space between plants:
45 cm (1½ ft)
Space between rows:
2 m (6 ft)

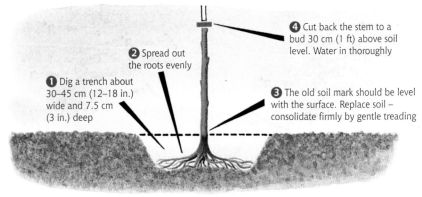

❶ Dig a trench about 30–45 cm (12–18 in.) wide and 7.5 cm (3 in.) deep

❷ Spread out the roots evenly

❸ The old soil mark should be level with the surface. Replace soil – consolidate firmly by gentle treading

❹ Cut back the stem to a bud 30 cm (1 ft) above soil level. Water in thoroughly

See page 10 for details of planting container-grown specimens. In this case there is no need to cut back the stem after planting.

# Support Systems

Summer-fruiting varieties must be supported. All sorts of systems can be seen in gardens – string or wire stretched between poles, chain-link fences, nylon twine across fruit cage supports, etc. For a very small garden the **single-post system** is useful, but harvesting is cumbersome and it is a poor method for more than a few plants. By far the most popular support is the **post-and-wire system** – canes are tied to the wires with soft twine. The **double-fence system** allows for more canes to be grown from each root – these canes are not tied to the wires.

## Post-and-wire system

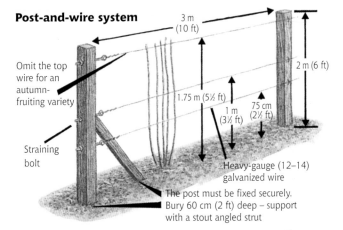

3 m (10 ft)

2 m (6 ft)

Omit the top wire for an autumn-fruiting variety

1.75 m (5½ ft)

1 m (3½ ft)

75 cm (2½ ft)

Straining bolt

Heavy-gauge (12–14) galvanized wire

The post must be fixed securely. Bury 60 cm (2 ft) deep – support with a stout angled strut

## Single-post system

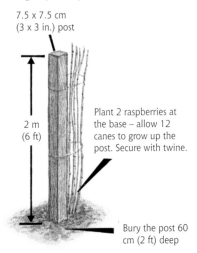

7.5 x 7.5 cm (3 x 3 in.) post

2 m (6 ft)

Plant 2 raspberries at the base – allow 12 canes to grow up the post. Secure with twine.

Bury the post 60 cm (2 ft) deep

## Double-fence system

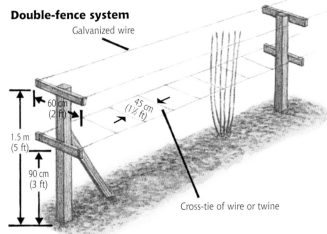

Galvanized wire

60 cm (2 ft)

45 cm (1½ ft)

1.5 m (5 ft)

90 cm (3 ft)

Cross-tie of wire or twine

# Pruning

### Newly planted raspberries

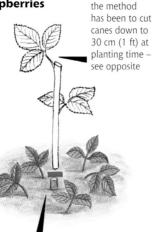

Traditionally, the method has been to cut canes down to 30 cm (1 ft) at planting time – see opposite

In spring cut down the old cane to near ground level when new growth appears

### Established summer-fruiting varieties

Immediately after picking cut down all the canes which have fruited. Retain the best 6–9 young unfruited canes and tie to wires 8–10 cm (3–4 in.) apart. In February cut back tall growth to 15 cm (6 in.) above the top wire

### Established autumn-fruiting varieties

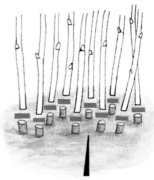

In February cut down all canes to ground level. Tie new canes to the wires with soft twine as they grow in the spring and summer. There is no need to thin these canes

A practice that is becoming more common with summer-fruiting varieties is to buy long-cane raspberries and plant using this method but don't cut back after planting. This allows side shoots to develop over the upper third of the canes, which will go on to produce fruit in their first year, rather than having to wait a year for cropping to start. During the first growing season the plants will also send up new canes from the base and these will go on to produce fruit in the second year.

## Seasonal Care

Raspberries are a thirsty crop. You must water them regularly during the first season if the weather is dry, and it is especially important to keep the soil moist when the fruit is swelling.

Keep weeds down by hoeing regularly – do not hoe too deeply or you will damage roots growing near the surface and this will promote suckering.

March is the usual time for mulching – see next column. Suckers should be removed in summer and stems growing away from the row should be pulled out.

With summer-fruiting varieties remove any flowers which may appear on the canes in the first summer after planting – this applies to container-grown plants, not long-cane ones.

Raspberry beetle is a serious pest. Spray when the first fruits start to turn pink – see page 107.

## Feeding and Mulching

Apply a general fertilizer such as Growmore at 60 g per sq. metre (2 oz per sq. yard) along the rows in March. Water in, and then apply a mulch. This should be a layer of well-rotted manure or compost – peat or bark is less effective. The purpose of this mulch is to keep the soil cool and moist in summer and to keep down weeds.

## Picking and Storage

Pick when the fruits are fully coloured but still firm. Pull each raspberry gently away from the stem, leaving the plug and stalk behind. Inspect the crop daily but try to pick fruit only when it is dry – wet fruit soon starts to go mouldy. Eat or freeze as soon as possible – small, slightly unripe fruits are the best ones for freezing.

## Raising New Plants

Raise new plants from your stock only if the canes look healthy and they were planted as certified stock within the last 2 seasons. Propagation is very simple – merely remove and plant suckers in October or November.

Lift up sucker and use secateurs to cut away from parent root. Plant immediately where the new raspberry is to be grown

# BLACKBERRIES AND HYBRID BERRIES

The blackberry is often regarded as a wild fruit (the bramble) for picking in late summer from the hedgerows. Not one for the garden – a rampant grower with viciously thorny stems. This generalization is no longer true – there are reasonably compact hybrids and there are several thornless types.

If you have a large trellis or a tall fence to clothe, then consider a modern blackberry variety. No soft fruit is less trouble – birds may take a share but protection is not necessary; flowers open between May and July so frost protection is not necessary. The long arching canes will yield 4.5–11 kg (10–25 lb) per plant and it will grow against a north-facing wall. Choose certified stock and the plant can remain productive for up to 20 years. Fruits appear between late July and September, depending on the variety. Grow *Oregon Thornless* or *Loch Ness* if people are likely to brush against the stems – choose *Ashton Cross* for real blackberry flavour or *Himalayan Giant* for a vandal-proof screen.

More and more people are turning to the hybrid berries which have been bred by crossing blackberries, raspberries, dewberries etc. There is a wide range of flavours here, and nearly all are less vigorous than the blackberry. Loganberries have been around for about a century and are the most popular, but tayberries are perhaps the best choice.

Blackberries and their hybrids need strong supports for their thick and strong stems. Training systems are described on page 80.

## Planting

Blackberries are not fussy about location – they will tolerate partial shade and also will grow in soils with less than efficient drainage. The canes are hardy but the tips can be damaged by hard frosts on exposed sites. Choose an area which has not grown raspberries or blackberries for a number of years.

Dig at least a month before planting – cultivate a 60 cm x 60 cm (2 ft x 2 ft) area at each planting site. All weeds must be removed at this stage. Remove the top spit of soil and add a 7.5 cm (3 in.) layer of garden compost at the bottom of the hole. Fork in and return the soil. Consolidate the surface and scatter a general fertilizer such as Growmore at 60 g per sq. metre (2 oz per sq. yard) over the ground before planting. Plant in the same way as raspberries – see page 74. Apply a mulch around the new canes in March.

Plant bare-rooted canes in November or December – plant in March if this is not possible.

Note: Container-grown specimens can be planted at any time of the year when the weather is suitable.

Space between plants:
1.8–4.5 m (6–15 ft)
(see Variety section)

Space between rows:
1.8 m (6 ft)

## Support Systems

There are a number of support systems. Thornless varieties can be grown over arches. All varieties can be grown against a series of strong wires stretched across fences, trellises or walls. The support has to hold last year's canes, which will bear fruit in summer or autumn, and also this year's canes for fruiting next year. The best support is the post-and-wire system – see illustration. This enables pruning, tying-in and picking to take place from both sides.

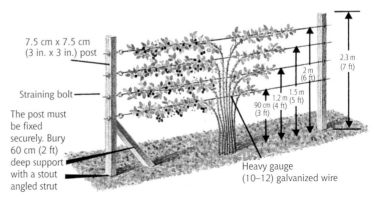

7.5 cm x 7.5 cm
(3 in. x 3 in.) post

Straining bolt

The post must be fixed securely. Bury 60 cm (2 ft) deep support with a stout angled strut

2.3 m (7 ft)

2 m (6 ft)

1.5 m (5 ft)

1.2 m (4 ft)

90 cm (3 ft)

Heavy gauge (10–12) galvanized wire

# Blackberry Varieties A–Z

### Ashton Cross

**Spacing:** 3.6 m (12 ft)

**Fruit size:** Medium

**Thorns:** Yes

**Picking time:** August–September

A wiry-stemmed variety which bears very heavy crops. The fruits are carried in large clusters and have a high reputation for freezing, but the feature noted in the catalogues is the 'true' flavour of the wild blackberry. You can expect to pick for 6–8 weeks.

### Black Butte

**Spacing:** 2.7 m (9 ft)

**Fruit size:** Very large

**Thorns:** Yes

**Picking time:** Early July– mid-August

This new blackberry from America has a vigorous and trailing habit, and is notable for its huge berries (twice the size of most blackberry fruits). The fruits are firm and juicy with a very good flavour and are well presented on strong laterals, making picking easier.

### Black Satin

**Spacing:** 3.6 m (12 ft)

**Fruit size:** Large

**Thorns:** No

**Picking time:** Late July–August

A fairly new one from the US – this one bears fruit earlier than any other thornless variety. It is not yet in the popular class but is well worth a trial. Growth is vigorous and the flavour of the fruit is rather sharp. Use it to cover a large arch or a pergola.

### Fantasia AGM

**Spacing:** 4.5 m (15 ft)

**Fruit size:** Very large

**Thorns:** Yes

**Picking time:** Late August– late September

A new star of the blackberry world – discovered on an allotment at Kingston-upon-Thames. Thorny, vigorous and cropping in August and September. Nothing special there – the unique features are the heavy yields and the outstanding size of the fruit. When well grown the berry is 2.5 cm (1 in.) in size.

### Himalayan Giant

**Spacing:** 4.5 m (15 ft)

**Fruit size:** Medium

**Thorns:** Yes

**Picking time:** Late August–late September

Choose this one only if you have a large space to cover and you require an impenetrable barrier or a windbreak. Very vigorous and very spiny – yields of 11 kg (25 lb) per plant are not uncommon. The flavour is fair and quite acid – for cooking rather than for eating fresh.

### John Innes

**Spacing:** 3 m (10 ft)

**Fruit size:** Very large

**Thorns:** Few

**Picking time:** September– October

Originally listed as a hybrid berry, really this is a blackberry bred at the John Innes Institute nearly 70 years ago. It is noted for its extra-large, jet-black, shiny fruits which have a distinct, sweet flavour. Spines are few and crops are heavy, but fruiting is rather too late to recommend it for northern districts.

### Loch Maree

**Spacing:** 1.8 m (6 ft)

**Fruit size:** Large

**Thorns:** No

**Picking time:** Late July– mid-August

One of a new generation of blackberries with a semi-erect, bush-like habit that can be grown without support. Attractive lilac-pink double flowers are followed by rounded black berries produced in clusters of 6–8 or more. An ideal garden variety.

### Loch Ness AGM

**Spacing:** 1.8 m (6 ft)

**Fruit size:** Large

**Thorns:** No

**Picking time:** Late August–September

Another blackberry 'bush', the stout stems are semi-erect, requiring little or no support. Winter-hardy with good yields and reasonably flavoured fruit, this one is bound to become very popular. Train and prune like a summer-fruiting raspberry.

### Merton Thornless

**Spacing:** 2.4m (8 ft)

**Fruit size:** Large

**Thorns:** No

**Picking time:** August–September

A fine thornless variety where space is limited – the rather short canes and the complete absence of thorns make maintenance a simple job. The yields are nothing special but the flavour is quite good. The plant is not as popular or as attractive as *Oregon Thornless* (see overleaf).

Ashton Cross

Black Butte

Fantasia

Himalaya Giant

**Oregon Thornless**

### Oregon Thornless

The popular thornless variety – the fruit has a mild flavour and the canes are easy to train. A decorative plant for a trellis or arch – the leaves are deeply divided and there are bright autumn colours. Other thornless varieties include *Smoothstem* and *Thornfree*.

**Spacing:** 3 m (10 ft)
**Fruit size:** Medium
**Thorns:** No
**Picking time:** Late August–late September

### Reuben

Bred in the USA, Reuben has been heralded as the first primocane blackberry, producing fruit on the current year's growth. The plant also has an upright growth habit, making it a far cry from the usual wild blackberry. The fruits are sweet and large, measuring up to 4.5 cm (1¾ in.) in length.

**Spacing:** 2.4–3 m (8–10 ft)
**Fruit size:** Very large
**Thorns:** Few
**Picking time:** Late July–late August

## Hybrid Varieties A–Z

**Boysenberry**

### Boysenberry

The fruit is round or oblong, looking like a large, dark and rather long raspberry. The flavour, however, is more akin to a wild blackberry. The prickly canes are borne in profusion – the *Thornless Boysenberry* is less vigorous. A good choice for sandy soils – boysenberry is remarkably drought-resistant.

**Parentage:** Loganberry x blackberry x raspberry
**Spacing:** 2.4 m (8 ft)
**Fruit size:** Large
**Skin colour:** Purple
**Picking time:** July–August

### Dewberry

The few catalogues which list this blackberry-like plant include it in the hybrid berry section, but it is not a hybrid at all. It is a distinct species – popular in the US but finding little favour in Britain. The slender stems can be left to trail over the plot as ground cover – the dull fruit has a good flavour.

**Parentage:** *Rubus caesius*
**Spacing:** 1.8 m (6 ft)
**Fruit size:** Small
**Skin colour:** Black, covered with grey bloom
**Picking time:** July

**Dewberry**

### Hildaberry

A new hybrid berry raised by an amateur and named for his wife. Not yet popular, but the flowers are 4–5 cm (1½–2 in.) across and the fruits can be more than 2.5 cm (1 in.) in diameter. The rounded berries can be eaten fresh or made into jam or jelly. You won't find this one at your garden centre – look through the specialist catalogues.

**Parentage:** Boysenberry x tayberry
**Spacing:** 2.4 m (8 ft)
**Fruit size:** Very large
**Skin colour:** Red
**Picking time:** June–mid-July

### Japanese Wineberry

An unusual type in many ways. First, it's not a hybrid – it is a separate *Rubus* species. It is also extremely decorative – the arching canes are covered with soft, bright red bristles. The small, seed-filled berries are sweet but bland, and they ripen all at once during August. Yields are quite low.

**Parentage:** *Rubus phoenicolasius*
**Spacing:** 1.8 m (6 ft)
**Fruit size:** Small
**Skin colour:** Orange or red
**Picking time:** August

**Japanese Wineberry**

### King's Acre

An old hybrid which is related to the loganberry but from which it differs in a number of ways. The red fruit turns black when ripe and the flavour is mild and similar to a blackberry – not sharp like a loganberry. It is also earlier ripening than the loganberry. Finally, it is suitable for a small garden.

**Parentage:** Raspberry x blackberry
**Spacing:** 2.4 m (8 ft)
**Fruit size:** Medium
**Skin colour:** Black
**Picking time:** Early July–August

### Loganberry

The basic hybrid bred in Canada in the last century was a prickly and moderately vigorous plant.

**Parentage:** Raspberry x blackberry
**Spacing:** 2.4 m (8 ft)
**Fruit size:** Large
**Skin colour:** Dark red
**Picking time:** Mid-July–August

### LY59 AGM

This is the type you can buy today. It is quite thorny, with a vigorous, blackberry-like trailing habit and long, raspberry-like fruits which have a sharper taste than a raspberry. Loganberries are excellent for culinary use but not for eating fresh.

**Loganberry**

**L654 AGM**

Originating from East Malling Research Station, this is a thornless and less vigorous variety. The fruit is dark red and longer than a raspberry, ripening over a long period. For the best flavour it should be allowed to ripen fully to a dark red colour before picking. A better choice for smaller gardens.

Marionberry

**Parentage:** Possibly blackberry 'Chehalem' × blackberry 'Olallie'

**Spacing:** 3.6 m (12 ft)

**Fruit size:** Large

**Skin colour:** Black

**Picking time:** Mid-July–September

## Marionberry

Something of a mystery – the marionberry has long been considered to be a hybrid berry, but is now thought to be a true blackberry, despite its sharp loganberry-like flavour. The canes are thorny and very vigorous. The most outstanding feature is the length of the picking season, extending for 8 or 9 weeks.

**Parentage:** Includes loganberry, youngberry, marionberry & boysenberry

**Spacing:** 4.5 m (15 ft)

**Fruit size:** Large

**Skin colour:** Dark red

**Picking time:** Early August–September

## Silvanberry

A recent introduction from Australia which is claimed to do well in heavy soils and exposed situations. It is also claimed that disease resistance is excellent, but it will have to stand the test of time. Fruits are large and sweet but growth is very vigorous and spreading – not one for a small garden.

Tayberry

**Parentage:** Raspberry x blackberry

**Spacing:** 4.5 m (15 ft)

**Fruit size:** Medium

**Skin colour:** Dark red

**Picking time:** Mid-July–late August

## Sunberry

A recent hybrid from East Malling – its main claim to fame is that the cropping period lasts for many weeks. It is far too vigorous and thorny to become really popular, although the very dark and glossy berries have a reasonably good loganberry-type flavour. Tayberry or tummelberry is a better choice.

**Parentage:** Raspberry x blackberry

**Spacing:** 2.4 m (8 ft)

**Fruit size:** Large

**Skin colour:** Dark purple

**Picking time:** Mid-July–August

## Tayberry AGM

Pick this one if you plan to buy only one hybrid berry, and ask for the virus-free Medana strain. Growth isn't rampant but yields are good and the juicy aromatic berries are larger than a loganberry and also much sweeter. The stems are slightly thorny and they can suffer in winter in exposed northern locations.

### Buckingham

The stems of *Buckingham* are thornless, but it is identical to the original tayberry in every other respect.

**Parentage:** Tayberry x tayberry seedling

**Spacing:** 2.4 m (8 ft)

**Fruit size:** Large

**Skin colour:** Red

**Picking time:** Mid-July–late August

## Tummelberry

A new hybrid berry from the Scottish Crop Research Institute – hardiness has been bred into the tayberry. Here is one which will stand up to cold winters, but some of the sweetness of the tayberry has been lost – the flavour is more like a loganberry. The hairy canes grow upright rather than arching.

Tummelberry

**Parentage:** Raspberry x blackberry

**Spacing:** 2.4 m (8 ft)

**Fruit size:** Large

**Skin colour:** Dark red

**Picking time:** August–September

## Veitchberry

One of the older raspberry/blackberry crosses. You will still find it in a number of catalogues because the fruit is larger than either parent and the flavour is excellent. Canes are stout and stiff rather than wiry and arching and the fruit is difficult to pull away from the plug. Birds can be a problem.

**Parentage:** Loganberry x dewberry

**Spacing:** 2.4 m (8 ft)

**Fruit size:** Large

**Skin colour:** Purple

**Picking time:** Late July–September

## Youngberry

The fruit looks like a boysenberry or a rounded loganberry. This hybrid has never become popular and you will have to search for a supplier. The flavour is very close to a loganberry and the juicy fruits contain very few seeds. If you want to try this one, choose the new *Thornless Youngberry* variety.

Youngberry

# Pruning and Training

The **fan** method of training gives high yields and is often recommended for the less vigorous hybrids. With this system the new canes are trained vertically and then along the top wire. The fruiting canes are tied as a fan to either side of this central column. After fruiting these old canes are removed and the central ones are then spread out to both sides to provide next year's fruiting canes. The fan method is a lot of work – even more troublesome is the **weaving** system which gives top yields. Here the new canes are woven up and down along the bottom 3 wires after the old canes have been removed in autumn. In summer the current wood is trained vertically and along the top wire.

The simplest, but not the most productive, system is **roping**. This is the procedure recommended for general garden use and is shown below. Wear leather gloves – burn old canes. In February cut back any dead shoot tips.

*Blackberries and hybrid berries are pruned after picking has finished. Take out the old fruit-bearing canes. Tie the strongest, new healthy canes along horizontal wires.*

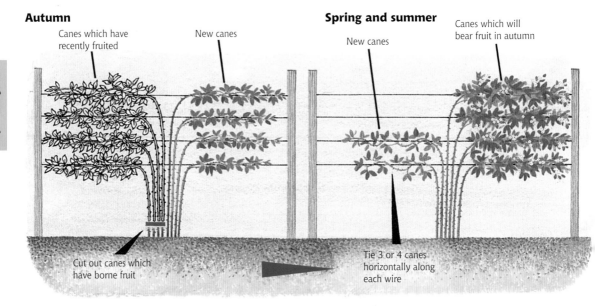

**Autumn**

Canes which have recently fruited

New canes

Cut out canes which have borne fruit

**Spring and summer**

New canes

Canes which will bear fruit in autumn

Tie 3 or 4 canes horizontally along each wire

# Seasonal Care

Water in summer if the weather is dry when the fruit has started to develop. Water the ground and not the stems in order to reduce the disease risk.

March is the usual time for mulching – see below. Canes growing away from the row should be pulled out. Keep weeds down by regular shallow hoeing.

Spray against raspberry beetle if this pest has been serious in previous years – see page 107

# Feeding and Mulching

Apply a general fertilizer such as Growmore at 60 g per sq. metre (2 oz per sq. yard) around each plant. Water in, and then apply a mulch. This should be a layer of well-rotted manure or compost – peat or bark is less effective. The purpose of this mulch is to keep the soil cool and moist in summer and to keep down weeds.

# Raising New Plants

**2** Use secateurs to sever 30 cm (1 ft) of the rooted tip from the parent stem. Plant where the new blackberry is to grow

**1** In July–September dig a hole and bury the tip of a healthy shoot about 15 cm (6 in) deep. Hold down with a hairpin – replace soil and tread down

# Picking and Storage

Pick when the fruits are fully coloured and soft. Pull each blackberry gently away from the stem – the plug generally comes away with the fruit. Pick when the fruit is dry – wet fruit soon starts to go mouldy. Eat or freeze as soon as possible – use in jams, jellies, pie-making, puddings, etc.

# BLACKCURRANTS

Blackcurrants are grown mainly for use in pies and puddings. They are also excellent for jam and jelly making, but are not popular for serving fresh. They are not difficult to grow, but you must know what to do at planting and pruning time. The reason is that most of the fruit is borne on last year's stems, and that means a regular supply of new growth from below ground level is required each year. To ensure this you must plant deeply with the root/shoot union well below the surface and you must cut out some old wood in winter.

Buy 2-year-old certified plants with at least 3 shoots. The bush should eventually reach a height and spread of 1.5 m (5 ft), producing 4.5–6.75 kg (10–15 lb) of fruit each summer and staying productive for 10–15 years. Reversion or big bud mite may reduce this long and active life by a few years, but the real threat until recently was frost. Older varieties bloom in early April, and both flowers and small fruitlets are susceptible to frost damage. Fortunately, plant breeders have introduced a range of late-flowering and partially frost-hardy types – the 'Ben' group and *Malling Jet* are examples. Choose *Ben Sarek* if the garden is small, *Ben Lomond* or *Ben More* as reliable all-rounders or *Malling Jet* for picking in August–early September. Some people claim that the modern varieties have lost the 'real' blackcurrant flavour, but there never has been a standard taste – *Baldwin* is distinctly sharp and early-fruiting *Laxton Giant* is sweet.

Soft Fruit
Blackcurrants

## Varieties A–Z

### Baldwin

**Bush size:** Medium
**Fruit size:** Medium
**Cropping:** Heavy
**Picking time:** Late July

The most popular choice until *Ben Lomond* came along. An age-old variety – choose the productive Hilltop strain. Growth is quite compact and the tart fruit hangs on the plant for a long time without splitting. The early flowering habit is a problem in northern districts.

### Ben Lomond

**Bush size:** Medium
**Fruit size:** Large
**Cropping:** Heavy
**Picking time:** Late July

The first of the 'Ben' varieties, which have brought late flowering, some mildew resistance and very heavy yields to blackcurrants. Growth is upright and reasonably compact – the strigs (flower stalks) are short and the plump berries have an acid flavour.

### Ben More

**Bush size:** Medium
**Fruit size:** Large
**Cropping:** Heavy
**Picking time:** Late July

A recent introduction to challenge *Ben Lomond*. The growth habit is neater and yields are reputed to be heavier. Best feature of all if you live in the north is the lateness of flowering – even later than *Malling Jet*. Resistance to mildew is quite high and the fruit flavour is sharp.

### Ben Nevis

**Bush size:** Large
**Fruit size:** Medium–large
**Cropping:** Heavy
**Picking time:** Late July

Similar to *Ben Lomond* in many ways – it flowers and fruits at the same time as *Ben Lomond*, the well-favoured berries are borne on short strigs and there is both mildew and frost resistance. The growth habit is different – it is taller, more vigorous and more upright.

Baldwin

Ben Nevis

81

Ben Sarek

Boskoop Giant

Jostaberry

Titania

### Ben Sarek

The main feature here is the dwarf nature of the bush – it grows only 1–1.2 m (3–4 ft) high and should be planted at 1.2 m (4 ft) spacings. There are inbred frost resistance and some mildew resistance, but also one or two drawbacks. Branch support is necessary and so is prompt picking before the fruits fall.

**Bush size:** Small
**Fruit size:** Large
**Cropping:** Heavy
**Picking time:** Mid-July

### Big Ben AGM

The large, shiny, black, sweet-flavoured fruits are more than double the size of standard varieties. Great for eating straight from the bush. Mature bushes develop a slightly spreading, arching habit and show good resistance to mildew and leaf spot. Ideal for the home garden.

**Bush size:** Large
**Fruit size:** Very large
**Cropping:** Heavy
**Picking time:** Early–mid-July

### Black Reward

A tall and vigorous Dutch variety launched to rival *Baldwin*. This one flowers 1–2 weeks later than the old favourite, which means much higher yields in summer following a cold spring. Flavour is good, but there is no mildew resistance. Spraying will be necessary.

**Bush size:** Large
**Fruit size:** Medium
**Cropping:** Heavy
**Picking time:** Late July

### Boskoop Giant

A giant of a plant – recommended spacing is 1.8 m (6 ft). Vigorous and spreading, but the yield is lower than that of a modern Ben variety. The sweet fruit appears early – provided that frost has not affected the cold-susceptible flowers. Not for the north or small gardens.

**Bush size:** Large
**Fruit size:** Large
**Cropping:** Moderate
**Picking time:** Early July

### Jostaberry

Not a blackcurrant variety – it's a blackcurrant x gooseberry hybrid. Grow it like a blackcurrant, but leave 1.8 m (6 ft) between the plants. Resistant to mildew and very prolific. You will tell the difference from an ordinary variety at picking time – the currants are twice the size.

**Bush size:** Large
**Fruit size:** Very large
**Cropping:** Heavy
**Picking time:** July

### Laxton Giant

Another early-fruiting giant like *Boskoop Giant*. Susceptible to spring frosts, but it is still chosen by exhibitors for the remarkable size of its fruit when well-grown. Plant 1.8 m (6 ft) apart – very vigorous and spreading. Berries are sweet and can be eaten fresh for dessert.

**Bush size:** Large
**Fruit size:** Very large
**Cropping:** Heavy
**Picking time:** Early July

### Malling Jet

Something different – a vigorous bush (too large for small gardens) which bears very long strigs of smallish berries in August and early September. Flowering is also very late so frost damage is unlikely. The flavour is acidic but rather bland. Plant 1.8 m (6 ft) apart.

**Bush size:** Large
**Fruit size:** Small–medium
**Cropping:** Moderate
**Picking time:** Late August

### Titania

This late-flowering Swedish variety has good frost resistance, making it a good choice for northern districts. The bush has a vigorous upright growth habit and is resistant to powdery mildew and rust. The very large berries have a high juice content and plenty of flavour.

**Bush size:** Medium
**Fruit size:** Large
**Cropping:** Heavy
**Picking time:** Early July–early August

# Planting

Blackcurrants will tolerate poor drainage better than other soft fruits, but they do need organic-rich soil and some shelter from the wind. Full sun is preferred but light shade is not a problem. Do not plant on a site prone to late frosts unless you grow a modern frost-resistant variety and are willing to cover the bushes if April nights are unusually cold.

Prepare the planting sites well before the bushes arrive – follow the instructions for blackberries on page 76.

Plant bare-rooted bushes in November – plant in February–March if this is not possible.
Note: Container-grown specimens can be planted at any time of the year when the weather is suitable.

Space between plants: 1.5 m (5 ft)
Space between rows: 1.5 m (5 ft)

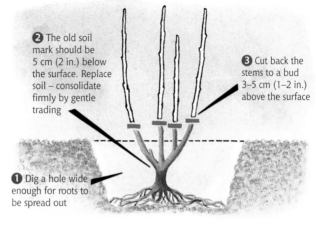

❷ The old soil mark should be 5 cm (2 in.) below the surface. Replace soil – consolidate firmly by gentle trading

❸ Cut back the stems to a bud 3–5 cm (1–2 in.) above the surface

❶ Dig a hole wide enough for roots to be spread out

# Pruning

Prune between November and March. Start 2 years after planting, cutting out weak branches. From 3 or 4 years onwards remove some old wood each year, as shown below.

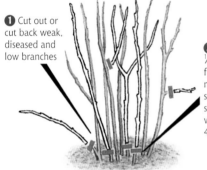

❶ Cut out or cut back weak, diseased and low branches

❷ Cut out about ⅓ or ¼ of the fruited branches to make room for new shoots. No wood should be retained which is more than 4 years old

# Raising New Plants

Take 25 cm (10 in.) cuttings from this season's wood in October – the cuttings should be pencil-thick. Make a sloping cut at the top – a straight cut at the bottom.

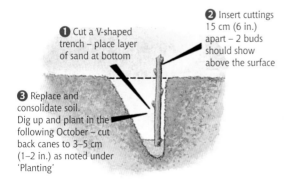

❷ Insert cuttings 15 cm (6 in.) apart – 2 buds should show above the surface

❶ Cut a V-shaped trench – place layer of sand at bottom

❸ Replace and consolidate soil. Dig up and plant in the following October – cut back canes to 3–5 cm (1–2 in.) as noted under 'Planting'

# Seasonal Care

In spring make sure that the bushes have not been lifted by frost – tread down if this has happened. Another spring job is to provide frost protection, especially for older varieties – drape netting or sacking over the bushes when night frosts threaten at flowering time. Remove the cover during the day.

Keep weeds under control. Hoe with great care – keep the blade near the surface to avoid damaging the shallow rooting system. It is better to keep weeds in check by mulching and hand pulling.

Water regularly and thoroughly during dry weather – keep water off the stems.

Bird protection is necessary. Drape nets over the bushes when the fruits begin to change from green to purple.

# Feeding and Mulching

Mulching is important. Place a mulch around the bushes after planting and renew this mulch each April with a 7–8 cm (3 in.) layer of garden compost or rotted manure.

Blackcurrants do require feeding. Spread about 120 g (4 oz) of Growmore around each bush in March before mulching and apply a potassium-rich liquid feed in summer when the fruits are swelling.

# Picking and Storage

Fruit is ripe and ready for picking about 7 days after it has turned blue-black. For immediate use you can pick individual currants – the fruits at the top of the strig are the first to ripen. For show purposes or for keeping in the refrigerator it is better to cut the whole strig. Harvest early varieties promptly.

Blackcurrants can be kept in the refrigerator for a week. For longer storage you should freeze, bottle or dry.

# GOOSEBERRIES

The gooseberry is a long-suffering plant – left unfed and unpruned a bush will continue to give some fruit for many years. Because of this it is sometimes called an 'easy' plant, but that is not true. Pruning in winter and summer is necessary to give an abundant crop which can be readily reached and which is not borne at ground level – and pruning among the vicious thorns is troublesome. Winter protection against birds is important in many areas, and failure to spray against American mildew can result in a totally mouldy crop.

Gooseberries bear a permanent framework of young and mature branches. The usual growth form is a 0.6–1.5 m (2–5 ft) high bush which is open centred and is borne on a 10–15 cm (4–6 in.) bare stem (leg). You can grow the plant as a cordon with 1, 2 or 3 upright 1.8 m (6 ft) stems – much less popular are the fan and the half-standard.

The bush yields 2.7–5.5 kg (6–12 lb) of berries; you can expect about 1–1.5 kg (2–3 lb) from a cordon. The plants are not generally troubled by virus and should remain productive for 10–15 years. Choose a 2- or 3-year-old bush for planting – cultivation is similar to the red or white currant.

Culinary varieties have a sharp taste and are not suitable for eating fresh. Dessert varieties are not usually available in the shops – they are larger, thinner-skinned and sweeter. The best plan is to buy a dessert variety from which you can pick unripe fruit in late May or June for cooking and then pick ripe fruit in July or August for eating fresh.

## Planting

The standard rules for siting soft fruit apply. Full sun is preferred – partial shade is tolerated. Protection from strong winds is necessary and frost pockets must be avoided.

Prepare the planting sites well before the bushes arrive – follow the instructions for blackberries on page 76.

### Bushes

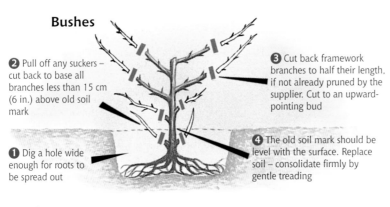

❷ Pull off any suckers – cut back to base all branches less than 15 cm (6 in.) above old soil mark

❸ Cut back framework branches to half their length, if not already pruned by the supplier. Cut to an upward-pointing bud

❶ Dig a hole wide enough for roots to be spread out

❹ The old soil mark should be level with the surface. Replace soil – consolidate firmly by gentle treading

Plant bare-rooted bushes in October–November or plant in February–March if this is not possible

Note: Container-grown specimens can be planted at any time of the year when the weather is suitable.

Space between plants:
1.5 m (5 ft)      (bushes)
45 cm (1½ ft) (single cordons)
60 cm (2 ft)    (double cordons)
90 cm (3 ft)    (triple cordons)

Space between rows: 1.5 m (5 ft)

### Cordons

As red and white currants – see page 91.

# Varieties A–Z

### Black Velvet

**Type:** Dessert
**Fruit size:** Small–medium
**Bush shape:** Upright
**Skin colour:** Dark red
**Picking time:** Mid-July

This is actually a gooseberry x Worcesterberry hybrid. The smallish grape-like fruits have a refreshing flavour – even more important is the complete resistance to American mildew of the bushes. It is an easy-to-grow plant which does well as a cordon – yields are high.

### Captivator

**Type:** Dessert/culinary
**Fruit size:** Small
**Bush shape:** Spreading
**Skin colour:** Dark red
**Picking time:** Mid-July

This Canadian variety forms a vigorous, spreading plant with almost spine-free stems, has good disease resistance and is very hardy. The small fruits are dark red and sweet, good for dessert when ripe or ideal for jams and pies if picked earlier.

### Careless AGM

**Type:** Culinary
**Fruit size:** Large
**Bush shape:** Spreading
**Skin colour:** Pale green
**Picking time:** Mid-July

The most popular variety – available everywhere. It is reliable and not fussy about soil type. The smooth berries are crisp – excellent for all cooking purposes. Susceptible to American mildew. Despite its popularity, *Whinham's Industry* or *Invicta* are better choices.

### Early Sulphur

**Type:** Dessert/culinary
**Fruit size:** Medium
**Bush shape:** Upright
**Skin colour:** Yellow
**Picking time:** Late June

An old favourite which is returning to the catalogues – it is one of the first to bear fruit. The primrose-yellow fruits are hairy and semi-transparent, and the flavour is good. It is a heavy cropper with a long-established reputation as a top gooseberry for jam-making.

### Golden Drop

**Type:** Dessert
**Fruit size:** Small
**Bush shape:** Upright
**Skin colour:** Greenish-yellow
**Picking time:** Mid-July

A good choice if you are looking for fruit to serve fresh – the small, round berries have thin skins and an excellent flavour. Another virtue where space is limited is the neat and compact growth habit. No resistance to American mildew. Yields are only moderate.

### Greenfinch AGM

**Type:** Culinary
**Fruit size:** Medium
**Bush shape:** Compact
**Skin colour:** Bright green
**Picking time:** Early July

The new, early variety from East Malling Research, giving heavy crops of good-quality smooth, green fruit. The bushes are less spiny than most others, with a compact habit. It has good resistance to foliar diseases. A good variety for organic growing, especially where space is limited.

### Hinnonmaki Green

**Type:** Dessert/culinary
**Fruit size:** Medium–large
**Bush shape:** Spreading
**Skin colour:** Olive green
**Picking time:** Mid-July

An excellent very hardy and vigorous mid-season variety, with a spreading habit and good disease resistance. A consistently prolific cropper of large, well-flavoured fruit, with a sweet aromatic taste. Does well in northern areas.

### Hinnonmaki Red

**Type:** Dessert/culinary
**Fruit size:** Medium
**Bush shape:** Upright
**Skin colour:** Dark red
**Picking time:** Mid-July

A leading variety across Europe, particularly in the colder areas. It produces heavy crops of large, sweet berries of excellent quality. It is a robust but slow-growing variety noted for its hardiness and disease resistance, so is ideal for small gardens and suitable for organic gardeners.

Careless

Early Sulphur

Green Gem

**Hinnonmaki Yellow**

**Invicta**

**Leveller**

### Hinnonmaki Yellow

A very hardy variety bred in Finland, producing heavy crops of fruits with an excellent, aromatic, apricot-like flavour. Bushes are rather more compact, but with spreading habit and good mildew resistance. Ideal for northern areas.

**Type:** Dessert
**Fruit size:** Medium–large
**Bush shape:** Spreading
**Skin colour:** Green-yellow
**Picking time:** Mid-July

### Invicta AGM

A real star of the gooseberry world. It is now widely grown and widely praised. The reasons are immunity to American mildew and very high yields. But it is extremely prickly, growth is vigorous and spreading, and the dessert quality when ripe is not particularly good.

**Type:** Dessert/culinary
**Fruit size:** Large
**Bush shape:** Spreading
**Skin colour:** Pale green
**Picking time:** Late July

### Jubilee

An improved form of the ever-popular *Careless*. All agree that it is more productive, and some catalogues claim all sorts of extra benefits. Examples are less spreading growth habit, better flavour as a dessert variety when ripe, earlier fruiting and larger-sized berries.

**Type:** Dessert/culinary
**Fruit size:** Large
**Bush shape:** Upright/spreading
**Skin colour:** Greenish-yellow
**Picking time:** Mid-July

### Keepsake

The fruit when ripe is almost white and is sometimes eaten fresh for dessert, but *Keepsake*'s main claim to fame is that the small unripe fruit can be picked in late May onwards for tarts, gooseberry fool, etc. The flavour is very good but late frosts can be a problem.

**Type:** Dessert/culinary
**Fruit size:** Medium–large
**Bush shape:** Spreading
**Skin colour:** Pale green
**Picking time:** Early July

### Lancashire Lad

This dual-purpose variety has some resistance to American mildew. The oval, hairy fruits have an attractive red colour and moderately good flavour, but it is disappointing where soil or situation is not good. Still, some people swear by it.

**Type:** Dessert/culinary
**Fruit size:** Medium–large
**Bush shape:** Spreading
**Skin colour:** Red
**Picking time:** Late July

### Langley Gage

The almost white berries are semi-transparent. They are smooth and oval, and are perhaps the sweetest of all gooseberries. Unfortunately there are not many suppliers but it is worth looking for. Yields can be high, but tend to be variable.

**Type:** Dessert
**Fruit size:** Small–medium
**Bush shape:** Upright
**Skin colour:** Pale yellow
**Picking time:** Late July

### Leveller AGM

Rightly popular, sharing the dessert crown with *Whinham's Industry*. The extra-large fruit is oval and downy – a standard choice for the exhibitor. Not just pretty – the flavour is exceptional. The catalogue will tell you that yields are very high – but only in good soil.

**Type:** Dessert/culinary
**Fruit size:** Very large
**Bush shape:** Spreading
**Skin colour:** Greenish-yellow
**Picking time:** Late July

### Lord Derby

Easy to recognize in July or August – a small bush with drooping branches and nearly black fruit. The round berries are smooth with an average to good flavour. Eye-catching, and excellent for the show bench. Yields are high for such a compact bush.

**Type:** Dessert/culinary
**Fruit size:** Very large
**Bush shape:** Arching
**Skin colour:** Dark red
**Picking time:** Early August

### Martlet

**Type:** Dessert/culinary
**Fruit size:** Large
**Bush shape:** Upright
**Skin colour:** Red
**Picking time:** Mid-July

This disease-resistant, vigorous plant has an upright growth habit, making it a far better-shaped bush which is much easier to maintain than many. The large, super-sweet fruits hang from the branches in clusters.

### May Duke

**Type:** Culinary
**Fruit size:** Medium
**Bush shape:** Upright
**Skin colour:** Dark red
**Picking time:** Early June

An old favourite which is traditionally picked at the green stage in May and June for pies and bottling and then in July at the red stage for stewing or even eating fresh. The fruit is oval and downy with a moderately good flavour – the bush is neat and compact.

### Pax

**Type:** Dessert
**Fruit size:** Medium–large
**Bush shape:** Spreading
**Skin colour:** Dark red
**Picking time:** Mid-July

The fruits are well shaped, slightly bristly, dark red, and of sweet dessert flavour when fully ripe. The plants are very vigorous with a spreading habit and the stems are virtually spineless. A good choice for beginner gardeners.

### Rokula

**Type:** Dessert
**Fruit size:** Medium
**Bush shape:** Spreading
**Skin colour:** Dark red
**Picking time:** Mid-July

This German-bred variety is very hardy and produces good crops of high-quality, very sweet-flavoured fruits early in the season. It is compact with a slightly drooping growth habit and has some mildew resistance. This variety grows well as a short standard, due to its drooping habit.

### Whinham's Industry AGM

**Type:** Dessert/culinary
**Fruit size:** Medium–large
**Bush shape:** Upright
**Skin colour:** Dark red
**Picking time:** Late July

A rival to *Leveller* as the best all-rounder. The fruit does not have *Leveller's* size or its sweetness, but the growth is upright and flavour is excellent. The main point is that it does much better than *Leveller* in shade and heavy soil.

### Whitesmith

**Type:** Dessert/culinary
**Fruit size:** Medium–large
**Bush shape:** Spreading
**Skin colour:** Yellowish-white
**Picking time:** Late July

A popular dual-purpose gooseberry which is a good choice. The near-white fruit is oval and downy with an exceptional flavour. The bush has a strong constitution, doing well in all sorts of soils. Yields are high. Available in fan and half-standard form.

### Worcesterberry

**Type:** Culinary
**Fruit size:** Small
**Bush shape:** Upright
**Skin colour:** Purple
**Picking time:** Late June

Small purple or black 'gooseberries' but it is really a currant species – *Ribes divaricatum*, from North America. Very vigorous and thorny – leave 2.4 m (8 ft) between the plants. Excellent for jam-making. Cultivate like a gooseberry – immune to American gooseberry mildew.

### Xenia

**Type:** Dessert
**Fruit size:** Medium–large
**Bush shape:** Upright
**Skin colour:** Red
**Picking time:** Late June

Originating from Switzerland, this variety produces heavy crops of red, sweet-flavoured, smooth-skinned berries early in the season. The strong, upright bushes have some resistance to powdery mildew. Stems are almost spine-free, making harvesting much easier.

May Duke

Pax

Rokula

# Pruning

## Bushes

Cut to an outward-facing bud for an upright variety or to an inward-facing bud for a spreading variety. Always wear a pair of leather gloves when pruning gooseberries.

The full method shown below is somewhat time-consuming. If growth is not vigorous, carry out the winter pruning and omit the summer cutting-back of the side-shoots.

### November–March

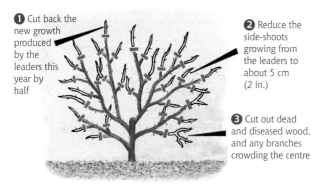

**1** Cut back the new growth produced by the leaders this year by half

**2** Reduce the side-shoots growing from the leaders to about 5 cm (2 in.)

**3** Cut out dead and diseased wood, and any branches crowding the centre

### Late June

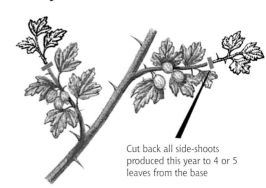

Cut back all side-shoots produced this year to 4 or 5 leaves from the base

## Cordons

As red and white currants – see page 91.

## Seasonal Care

Winter protection against birds is necessary – finches attack the buds which produce next year's fruit. Netting is suitable, but gooseberries are best grown in a fruit cage.

Check the plants after a heavy frost – firm down with your feet if the bushes have been lifted. Flowering time is early or mid-April – cover at night with netting or sacking if frosts threaten at this time.

Keep weeds down by mulching and by hand pulling. Hoeing can damage surface roots and lead to excess suckering. Pull suckers away from roots during the dormant season.

Water thoroughly and regularly in summer if the weather is dry and the fruit is beginning to swell.

If the crop is heavy, start to thin the fruits in May or June. Remove some of the berries when they are large enough for cooking – do this in stages until the fruits which are to mature to full ripeness are left 5 cm (2 in.) apart. Cover the ripening fruit with netting.

## Feeding and Mulching

Apply a general fertilizer such as Growmore at 60 g per sq. metre (2 oz per sq. yard) around each bush or cordon in March. Water in, and then apply a mulch of well-rotted manure or compost to keep down the weeds. Use peat or bark if compost is not available.

## Picking and Storage

The usual plan is to pick some small green fruit from both culinary and dessert varieties in late May. These are crisp and tart – excellent for crumbles, pies, tarts, etc. For dessert varieties which are to be eaten fresh, you will have to wait for the fully ripe stage. The fruit will be fully coloured and rather soft when gently squeezed. Not all will ripen at once – go over the bushes 2 or 3 times. Eat as soon as possible, although ripe gooseberries can be kept in a refrigerator for a couple of weeks.

## Raising New Plants

Take 30 cm (12 in.) cuttings from this season's wood in October – the wood should be well-ripened. Make a sloping cut at the top – a straight cut at the bottom.

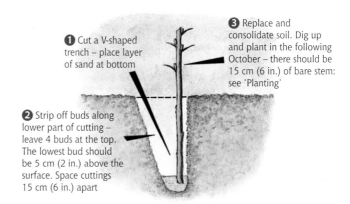

**1** Cut a V-shaped trench – place layer of sand at bottom

**2** Strip off buds along lower part of cutting – leave 4 buds at the top. The lowest bud should be 5 cm (2 in.) above the surface. Space cuttings 15 cm (6 in.) apart

**3** Replace and consolidate soil. Dig up and plant in the following October – there should be 15 cm (6 in.) of bare stem: see 'Planting'

# Red and White Currants

Red and white currants are becoming more common in supermarkets and farm shops thanks to an increased interest in cooking. However, they can be expensive, so growing your own ensures a midsummer supply of these delicious fruits without breaking the bank.

Red currants are tart – use them for pies, jams, jellies and wine-making. The white currant is a sport of the red one – the flavour is grape-like and the fruits are perfectly suitable for serving fresh.

The big surprise is that these two plants are planted, pruned and cared for like gooseberries and not like blackcurrants to which they are closely related. The usual form is an open-centred, goblet-shaped bush with a height and spread of about 1.5–1.8 m (5–6 ft). Fruit is carried on stubby side-shoots on the main branches – you can expect about 4.5 kg (10 lb) and the plant should remain productive for about 15 years. The purpose of annual pruning is to ensure an abundance of these fruiting spurs.

Red and white currants can also be grown as vertical cordons bearing 1, 2 or 3 stems. Stretch wires horizontally across a wall or fence 60 cm (2 ft) and 1.2 m (4 ft) above the ground – fix firm canes where the cordons are to be planted (see page 91). You can expect about 1.3 kg (3 lb) of fruit from each single cordon.

Buy 2-year-old plants from a reputable supplier – look for a stout bush with at least 4 evenly spaced branches. The choice of varieties is not large – the usual red currant choice is *Red Lake* or *Jonkheer van Tets*, although *Redstart* is expected to be the variety of the future. *White Versailles* is a good all-rounder and is the white currant you are most likely to find available at the garden centre.

## Varieties A–Z

### Jonkheer van Tets AGM

**Type:** Red currant
**Fruit size:** Large
**Bush shape:** Upright
**Picking time:** Early July

This one starts the red currant picking season. The bright fruit is juicy and rated very highly for flavour. Crops are heavy and bushes grow large with untidy and brittle branches. This variety does best grown as a cordon. Quite widely available.

### Red Lake AGM

**Type:** Red currant
**Fruit size:** Very large
**Bush shape:** Upright
**Picking time:** Late July

This reliable variety remains the most popular all-rounder. The fruit is carried on long trusses which are easy to pick, and heavy crops are borne on moderately vigorous bushes. No problems – flavour is good, juice content is high and stocks are sold everywhere.

### Redpoll

**Type:** Red currant
**Fruit size:** Medium–large
**Bush shape:** Spreading
**Picking time:** Mid–late August

A new variety from East Malling Research which produces very long strigs of bright red, late-ripening fruits. This vigorous variety has a spreading habit and moderate resistance to leaf spot and American gooseberry mildew. Suitable for growing as a cordon.

Red Lake

Redpoll

89

**Redstart**

**Rovada**

**Blanca**

**White Versailles**

### Redstart

This newer variety is slow growing, a good choice if space is limited. It is very late, cropping in mid-August. The bright fruits are produced in long strigs and yields are reported to be high and consistent. The flavour is distinctly acid.

**Type:** Red currant
**Fruit size:** Medium
**Bush shape:** Upright
**Picking time:** August

### Rondom

A dark red variety – the acid, rather dry fruits are borne in tight trusses. Crops are heavy and growth is vigorous – recommended for exposed sites where wind can damage other varieties. Growth is upright at first, but mature bushes have a spreading habit.

**Type:** Red currant
**Fruit size:** Medium
**Bush shape:** Upright/spreading
**Picking time:** Early August

### Rovada

A late-flowering and late-fruiting Dutch variety, with an erect habit which has become increasingly popular with gardeners. Most of its other features are typical of many red currants, but its cropping capacity appears to be exceptional. Flavour is good.

**Type:** Red currant
**Fruit size:** Large
**Bush shape:** Upright
**Picking time:** Early August

### Stanza AGM

A newer red currant with some interesting features. The fruits are quite small and darker than the popular varieties – it also flowers later which means that the chance of frost damage is lessened. Flavour is more acid and the growth habit is more compact than some varieties.

**Type:** Red currant
**Fruit size:** Small–medium
**Bush shape:** Upright
**Picking time:** Late July

### Blanca

This new Dutch-bred variety is one of the highest-yielding white currants, producing long strigs of pearly-white, large, sweet-flavoured berries. Heavy-cropping and vigorous with a spreading habit, it is becoming increasingly popular.

**Type:** White currant
**Fruit size:** Large
**Bush shape:** Spreading
**Picking time:** Late July–August

### White Dutch

Long trusses of pale golden fruits are borne in midsummer. Flavour is not as good as *White Grape*, and the growth habit is more spreading and untidy than the other white currant varieties. It scores by producing heavier crops than many of the others.

**Type:** White currant
**Fruit size:** Large
**Bush shape:** Spreading
**Picking time:** Late July

### White Grape AGM

Second to *White Versailles* in popularity. It is still not easy to find a supplier – go to a specialist fruit nursery. The sweet flavour of *White Grape* is excellent and unmatched by *White Versailles*, but the yield is lower. Colour is near-white.

**Type:** White currant
**Fruit size:** Large
**Bush shape:** Upright
**Picking time:** Mid-July

### White Versailles

The first white currant in both popularity and picking time. The trusses are long and heavy – the fruits are pale yellow and pleasantly sweet. It is thoroughly reliable, giving good crops year after year. It is the variety to buy if you plan to grow just one white currant.

**Type:** White currant
**Fruit size:** Large
**Bush shape:** Upright
**Picking time:** Early July

# Planting

Choose a sheltered spot in sun or semi-shade. Any reasonable garden soil will do, but poor drainage will cause problems. Avoid a frost pocket.

Prepare the planting sites well before the bushes arrive – follow the instructions for blackberries on page 76.

### Bushes

As gooseberries – see page 84.

### Cordons

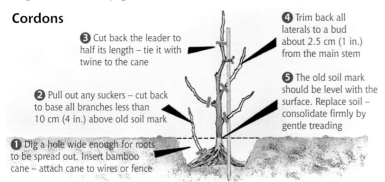

**①** Dig a hole wide enough for roots to be spread out. Insert bamboo cane – attach cane to wires or fence

**②** Pull out any suckers – cut back to base all branches less than 10 cm (4 in.) above old soil mark

**③** Cut back the leader to half its length – tie it with twine to the cane

**④** Trim back all laterals to a bud about 2.5 cm (1 in.) from the main stem

**⑤** The old soil mark should be level with the surface. Replace soil – consolidate firmly by gentle treading

Plant bare-rooted bushes in October–November or plant in February–March if this is not possible.

Note: Container-grown specimens can be planted at any time of the year when the weather is suitable.

Space between plants:

| | |
|---|---|
| 1.5 m (5 ft) | (bushes) |
| 45 cm (1½ ft) | (single cordons) |
| 1 m (3 ft) | (double cordons) |
| 1.2 m (4 ft) | (triple cordons) |

Space between rows: 1.5 m (5 ft)

*Summer pruning allows more light and air into the centre of the bush.*

# Pruning

### Bushes

As gooseberries – see page 88.

### Cordons

#### February

Continue basic procedure carried out at planting time. Prune back the leaders to leave about 15 cm (6 in.) of this year's growth. Cut back all laterals to a bud about 2.5 cm (1 in.) from the main stem.

When the top of the cane is reached it will be necessary to cut back the leader to just one bud above last year's growth.

#### Late June

Cut back all side-shoots produced this year to 4 or 5 leaves from the base – see gooseberries (page 88). This gets rid of aphids clustered at the shoot tips and also reduces the risk of disease.

As it extends, tie the leading shoot to the cane.

Single cordon   Double cordon   Triple cordon

# Seasonal Care

Birds in winter are a problem – finches attack the buds. Red and white currants are best grown in a fruit cage – if this is not possible it will be necessary to cover the plants with netting.

Check the plants after a heavy frost – firm down with your feet if the bushes have been lifted. Flowering time is between late March and the end of April – cover at night with netting or sacking if frosts threaten at this time.

Keep weeds down by mulching or hand pulling. Hoeing can damage the mass of surface roots and lead to excess suckering. Pull suckers away from the roots during the dormant season.

Water thoroughly and regularly in summer if the weather is dry and the fruit is beginning to swell.

Birds find the small and succulent fruit irresistible. It will be necessary to cover the plants with netting when the ripening currants begin to show colour.

# Feeding and Mulching

Apply a general fertilizer such as Growmore at 60 g per sq. metre (2 oz per sq. yard) around each bush or cordon in March. Water in, and then apply a mulch of well-rotted manure or compost.

# Picking and Storage

Pick when the fruit is fully coloured and shiny – a dull surface indicates you have waited too long. Not all will ripen at once – go over the bushes 2 or 3 times. Pick the whole cluster to avoid damaging the fruits – do not pick off the currants singly. Use as soon as possible, although red and white currants can be kept in the refrigerator for about a week.

# Raising New Plants

As gooseberries – see page 88.

# HEATHLAND BERRIES

All the heathland berries have one fundamental feature in common – they must have moist and distinctly acid soil. Providing this need is met they are easy to grow – pruning is simple and there is no need to spray with insecticides or fungicides.

North America is the home of the heathland berry, but they are increasing in popularity in the shops and gardens in Britain and have now started to appear in most nursery catalogues. There are three types. The **Lowbush blueberry** grows wild in boggy areas and is very rarely cultivated. The **Highbush blueberry** is the star of the group – very popular in the US and now fairly readily available from nurseries in Britain. Try it by all means, but your soil will have to be less than pH 5.5. The **cranberry** is even more demanding – the soil has to be less than pH 4.5.

The only one worth trying, especially for the beginner, is the Highbush blueberry. Have a go if your soil grows rhododendrons and azaleas *really* well with no sign of leaf yellowing – otherwise you should plant the specimens in stout containers. These will need regular watering with soft water or rainwater – a chore if you live in a hard-water area. Still, the result is worthwhile. Dense clusters of pinkish flowers in spring, succulent fruit in midsummer and brilliant red foliage in autumn.

## Varieties A–Z

### Cranberry

Unless your garden is a bog there is only one practical way to grow cranberries. Dig a hole about 1 m (1 yard) square and 1 spit deep. Line with polythene and fill with a mixture of 3 parts moss peat, 1 part loam, 1 part sharp sand, 1 part bark chips. Add 3 handfuls of bone meal to each barrow load. Plant in autumn or spring with 30 cm (1 ft) between plants. Water regularly with soft water and pick the red berries when they start to soften.

### Highbush blueberry

There are several ways of growing the Highbush blueberry in the garden. The obvious place is the fruit plot. Free-draining soil is necessary – if it is not gritty, acid and rich in organic matter you should fill each planting hole with a mixture of moss peat, coarse sand, soil and a little sawdust.

The bush will eventually have a height and spread of about 1.5 m (5 ft) and will bear 2.5 kg (6 lb) of fruit – dark blue berries about 1.25 cm (½ in.) across with a grey bloom. Plant 1.5 m (5 ft) apart in autumn or spring. Buy 2- or 3-year-old container-grown specimens. The site should be reasonably sunny and sheltered from strong winds. Frost is rarely a problem.

Prune in winter. Cut out damaged or dead branches and a few old stems which have borne fruit. In spring sprinkle a little fish, blood and bone fertilizer around plants and mulch with well-rotted organic matter, leaf mould or shredded bark. Water regularly with soft water in dry weather.

If your soil is acid it is quite practical to plant blueberries in a rhododendron border – the bright autumn tints provide a welcome splash of colour at the end of the season. If, however, your soil is quite definitely not acid then it is far better to plant them in stout containers which should be at least 45 cm (1½ ft) deep. Use an ericaceous compost or make your own by mixing 75% moss peat and 25% shredded bark with a small amount of sulphate of ammonia, sulphate of potash and bone meal.

Birds can be a problem – cover the bush with netting as the fruits begin to ripen. When they begin to soften and have been blue for a week, pick individual berries – you will have to go over the bush several times as not all the berries ripen at once. The fruits can be stored in the refrigerator for 2–3 weeks. Use in pies or tarts – best of all, find a recipe for blueberry muffins.

Several varieties are available (right), but for maximum yields grow 2 varieties to ensure cross-pollination.

### Lowbush blueberry

The bilberry (other names: whortleberry, blaeberry) is a fruit of the countryside. It grows about 30–60 cm (1–2 ft) high and is a common sight in high moorland in the north. The fruits appear in autumn – sweet, near-black berries with a grey bloom. Eat them fresh or use in pies.

### Bluecrop

This is considered the best all-round variety, producing consistent yields of large, high-quality berries with a slightly tart flavour. It has good disease resistance. There are other more productive varieties, but none better in the garden.

**Fruit size:** Medium–large

**Bush shape:** Upright/spreading

**Skin colour:** Medium blue

**Picking time:** August

### Chandler

Produces the largest blueberries of any variety, sometimes the size of cherries, with an excellent medium-sweet flavour. This variety has a very long cropping season lasting up to 6 weeks, producing consistently high yields.

**Fruit size:** Very large

**Bush shape:** Rounded/spreading

**Skin colour:** Dark blue

**Picking time:** Early August–mid-September

### Earliblue

This vigorous bush is popular as the earliest ripening variety, meaning it usefully extends the fruiting season. This disease-resistant variety matures early and produces large, sweet berries year after year on long, arching stems.

**Fruit size:** Medium–large

**Bush shape:** Open/upright

**Skin colour:** Light blue

**Picking time:** Early July

### Herbert

A less vigorous grower, producing large fruits of exceptional quality. The very dark, lightly speckled fruits are considered to be the best flavoured of all varieties. Not the heaviest cropper, but good for domestic gardens.

**Fruit size:** Large

**Bush shape:** Open/upright

**Skin colour:** Dark blue

**Picking time:** Mid–late August

### Spartan

A prolific bearer of good-quality fruits which ripen over a long period. *Spartan* is an early-cropping blueberry ripening in early to mid-July. It has berries with a delicious, tangy, sweet flavour. The upright habit makes picking easier.

**Fruit size:** Large

**Bush shape:** Upright/bushy

**Skin colour:** Light blue

**Picking time:** Early–mid-July

# GRAPES

You will read in all the books that outdoor grapes were grown extensively in Roman Britain – what isn't said is that their decline was due to a change in climate rather than a change in our taste for wine. For hundreds of years the British climate was too cold for the grapevine, but now you can expect successful results south of a line stretching from Gloucester to the Wash.

You can of course grow grapes under glass both inside and outside this area, but you should consider carefully whether you are prepared to devote space and time to this vigorous crop which demands attention nearly all year round.

A mature grapevine will yield about 6.75 kg (15 lb) of fruit – enough to make 4.5 litres (1 gallon) of wine. It should stay productive for about 40 years, but you can't expect any fruit until 3 years after planting. Training and regular pruning are essential if you are to avoid a plant with masses of foliage and very few bunches of fruit – there are several different systems of training and pruning and none can be truly described as simple. Grapes, then, are for the enthusiast and not the reluctant gardener.

## Outdoor Grapes

**Black types**
purple, maroon

**White types**
green, gold or pale pink

## Greenhouse Grapes

**Black types**
purple, maroon

**White types**
green, gold or pale pink

Outdoor varieties are generally smaller than the greenhouse ones, and the prominent surface bloom is usually missing.

Outdoor varieties are often grown for wine-making as there is rarely enough sunshine to develop sweet and juicy flesh. There are, however, several dessert varieties which can be used as table grapes for eating fresh. A few types are recommended for both outdoor and greenhouse use – *Leon Millot*, *Chasselas* and *Seyval* are examples.

Early varieties are ready for picking in September. You will have to wait until the first half of October for mid-season varieties and until the second half of October for the late ones. Choose an early variety if you live close to the north–south dividing line.

White types are generally more successful outdoors than black.

The range of varieties for greenhouse culture is greater and so is the chance of success. Many will grow in a cold house, but some need a minimum of 12–13°C (55°F) in early spring.

The Sweetwater group, including the ever-popular *Black Hamburgh*, are easiest to grow. Thin-skinned and very sweet, they need no heat and ripen early. Cut as soon as they reach the recommended stage. The Muscat group have the best flavour but are more trouble, calling for some heat and for hand pollination. They can be left on the vine until they develop a slightly raisin-like taste. The Vinous group provide grapes for Christmas or even later. They hang on the vine for months, but the house must be kept warm throughout late autumn and early winter.

# Planting

Grapes need as much sun as possible outdoors – to avoid disappointment choose a wall which faces south, south-east or south-west. Good drainage is vital – don't bother to plant this crop if waterlogging is a problem. All soils but clays and shallow chalks can be used, but the best types are gritty or sandy soils enriched with organic matter.

Dig at least a month before planting – cultivate a 60 cm x 60 cm (2 ft x 2 ft) area at each planting site. All weeds should be removed at this stage. Remove the top spit and add a layer of garden compost at the bottom of the hole. Fork in and return the soil. Consolidate the surface and scatter a general fertilizer such as Growmore at 60 g per sq. metre (2 oz per sq. yard) over the area before planting.

Plant in October–November or February–March. March is the best time to plant outdoors – cover winter-planted specimens with a 10 cm (4 in.) layer of compost or leaf mould.

Space between plants: 1.5 m (5 ft)

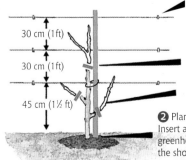

30 cm (1ft)

30 cm (1ft)

45 cm (1½ ft)

❶ Stretch support wire 30 cm (1 ft) apart between vine eyes set 23 cm (9 in.) away from the wall (outdoors) or supports (greenhouse)

❸ Cut back the main shoot ('rod') to 60 cm (2 ft). Tie the rod to the cane support

❹ Cut back all side-shoots to 1 bud

❷ Plant the specimen firmly to the old soil mark. Insert a cane before planting. Some people plant greenhouse grapes outside the house and bring the shoot inside through a hole at the base

Brandt

Lakemont Seedless

Seyval

# Outdoor Varieties
## A–Z

### Bacchus

Bred in Germany, this is one of the better grape varieties for cooler climates. A fine-quality white grape with a distinct flavour and good habit. It is a reliable, heavy-cropping and strong-growing variety, and is easy to grow.

**Type:** Dessert/wine
**Fruit size:** Medium
**Skin colour:** Greenish-white
**Picking time:** Mid-season

### Brandt

Both fruit and bunches are small, but this hybrid is very popular and easy to find in the catalogues. The heavy crop is sweet and growth is vigorous – useful for covering an unsightly wall. Some mildew resistance. A warm and sunny spot is essential.

**Type:** Dessert
**Fruit size:** Small
**Skin colour:** Purple
**Picking time:** Late

### Cascade

Sometimes listed as *Seibel 13.053*. The small grapes borne in tight clusters produce a dark, acid juice – the red wine quality is described as fair. Resistance to mildew is a definite advantage, as is its vigour if an effective screen is required.

**Type:** Wine-making
**Fruit size:** Small
**Skin colour:** Dark purple
**Picking time:** Mid-season

### Lakemont Seedless

This seedless grape should thrive in all areas except for the far north and is also suitable for indoor cultivation. Produces large bunches of pale yellow-green fruits, with a good, sweet flavour and sets evenly. Has some resistance to powdery and downy mildew.

**Type:** Dessert
**Fruit size:** Medium
**Skin colour:** Yellow
**Picking time:** Early–mid-season

### Leon Millot

A vigorous French variety, producing near-black grapes for the table or for turning into a fair-quality red wine. Bunches are small – resistance to mildew is good. Can be grown under glass in northern districts. Cropping is both good and reliable.

**Type:** Dessert/wine-making
**Fruit size:** Medium
**Skin colour:** Blue-black
**Picking time:** Mid-season

### Müller Thurgau

Our most popular grape for wine-making – sometimes sold as *Riesling Sylvaner*. The hock-type product has a delicate flavour and fine bouquet. The grapes are good but not outstanding for eating fresh. Yields are heavy, but not if the weather is poor at pollination time.

**Type:** Dessert/wine-making
**Fruit size:** Small–medium
**Skin colour:** Golden
**Picking time:** Mid-season

### Regent

A fairly vigorous, disease-resistant, outdoor grape that will produce clusters of large, dark red fruits, which mature to a glossy black in warm summers. The fruits have a sweet flavour and the leaves turn a stunning red before falling in the autumn.

**Type:** Dessert/wine-making
**Fruit size:** Large
**Skin colour:** Dark red/black
**Picking time:** Mid-season

### Seyval

A popular wine-making grape, sometimes sold as *Seyval Blanc* or *Seyve Villard*. Easy and reliable – often recommended for chalky soils and colder areas. The wine is described as light and fruity. There is some resistance to mildew.

**Type:** Wine-making
**Fruit size:** Medium
**Skin colour:** Green
**Picking time:** Mid-season

# Greenhouse Varieties A–Z

### Alicante

**Type:** Vinous

**Fruit size:** Large

**Skin colour:** Black with blue-grey bloom

**Picking time:** Late

A good choice for table or show bench – sometimes called *Black Tokay*. The bunches are large but lop-sided – thinning is necessary. It sets fruit very freely, crops are heavy and flavour is very good – but it is a Vinous grape and so heat is required.

Alicante

### Black Hamburgh

**Type:** Sweetwater

**Fruit size:** Large

**Skin colour:** Dark red or purple with blue bloom

**Picking time:** Early

By far the most popular greenhouse grape, and quite rightly so. It is the easiest to grow, and the flavour of the round succulent grapes is good. It sets freely, ripens well without heat and consistently provides large bunches year after year.

### Buckland Sweetwater

**Type:** Sweetwater

**Fruit size:** Large

**Skin colour:** Amber

**Picking time:** Early

A compact vine – worth considering if space is limited. Generally trouble-free – it sets freely and yields are good. The round, pale green fruits turn pale golden when ripe and the flavour is sweet but not outstanding. Regular feeding is necessary.

### Chasselas

**Type:** Sweetwater

**Fruit size:** Medium

**Skin colour:** Pink or amber

**Picking time:** Early

This variety has been around for centuries, and the various forms appear in the catalogues as *Chasselas d'Or*, *Royal Muscadine*, *Chasselas Rosé Royale*, etc. Easy, reliable and high yielding like *Black Hamburgh*. Sometimes recommended for outside wall culture.

Black Hamburgh

### Flame

**Type:** Sweetwater

**Fruit size:** Medium

**Skin colour:** Dark red

**Picking time:** Mid-season

A heavy-bearing table grape cultivar from California, producing seedless fruits which have a crunchy texture and are full of juice. This variety is best grown in an unheated greenhouse or conservatory. The grapes should be picked when ripe, as they do not ripen after harvest.

### Madresfield Court

**Type:** Muscat

**Fruit size:** Large

**Skin colour:** Near-black with blue-grey bloom

**Picking time:** Early

A very high-quality black variety – the oval grapes are juicy and full of flavour but the skin is tough. The bunches are long and wide and you must pick them as soon as they are ripe as the grapes tend to split if left on the vine.

### Muscat Hamburgh

**Type:** Muscat

**Fruit size:** Large

**Skin colour:** Dark red or purple with blue bloom

**Picking time:** Late

The flavour of the fresh fruit is excellent and so is the wine it produces, but it must not be confused with the popular and easy-to-grow *Black Hamburgh*. This one needs heat. Cross-pollination with another variety is recommended.

### Muscat of Alexandria

**Type:** Muscat

**Fruit size:** Large

**Skin colour:** Amber with white bloom

**Picking time:** Late

This is not a high-yielding variety and you will have to provide some heat either in spring or autumn. But the quality is excellent – large bunches of oval fruit which taste as good as or better than any other home-grown grape. It can be unreliable.

Chasselas

# Pruning and Training

There are several methods – the cordon system is the usual one for greenhouse culture and the Guyot system is sometimes recommended for outdoor vines. For the sake of simplicity the cordon system is recommended here for both greenhouse and outdoor cultivation.

### 1st and 2nd year after planting

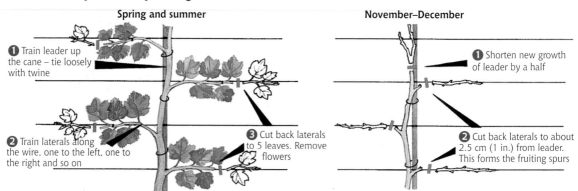

**Spring and summer**

❶ Train leader up the cane – tie loosely with twine

❷ Train laterals along the wire, one to the left, one to the right and so on

❸ Cut back laterals to 5 leaves. Remove flowers

**November–December**

❶ Shorten new growth of leader by a half

❷ Cut back laterals to about 2.5 cm (1 in.) from leader. This forms the fruiting spurs

### 3rd and subsequent years after planting

Repeat the basic routine outlined above with the following modifications. When the flowers appear in spring, cut back the fruiting lateral to 2 leaves beyond the truss – cut back sub-laterals to just 1 leaf. Keep only 1 fruiting lateral per spur – rub out all the others. For the first few years leave just 1 bunch per lateral – on a mature vine 2 or 3 bunches can be grown on each lateral.

In November or December prune the leader ('rod') as above. When the top wire is reached prune it back to 2.5 cm (1 in.) beyond last year's growth – in other words, treat it as a lateral.

## Seasonal Care under Glass

In midwinter remove loose bark from the pruned rod and paint with a winter wash – clean the glass behind the vine. This usually involves undoing the upper ties and letting down the rod behind before retying in the spring.

In early spring apply a general fertilizer such as Growmore at 60 g per sq. metre (2 oz per sq. yard) around the base of the plant. Water the surface and spread a 5 cm (2 in.) mulch of well-rotted compost around the stem. In March it is time to turn on the heat if available – start gently at first and increase to 12–16°C (55–60°F) after a few weeks. In an unheated house growth begins in April – the ventilators should be closed and damping down can begin.

In May it will be necessary to promote fertilization by shaking the stem at midday or by hand pollination (see page 43).

Watering will be necessary once active growth starts. Thoroughly soak the ground around the base of the stem every 3 weeks – apply a high-potash liquid fertilizer according to the maker's instructions from the time the first fruits form until the grapes start to colour.

Thinning will be necessary if the set has been heavy. Begin when the berries are pea-sized, using a pair of long-bladed scissors in one hand and a small forked stick in the other. Never touch the fruit with your fingers – remove some of the berries within the bunch and get rid of small berries. Do this on several occasions over 7–10 days so that there is a 12 mm (½ in.) space between the remaining grapes. A fussy job – but worth it for top-quality results. Snip off any diseased fruits during this ripening stage.

Prune the laterals as noted above during the spring and summer months and apply shading from June until late August. In August reduce ventilation and stop damping down in order to aid ripening. Reduce watering at this stage.

## Seasonal Care Outdoors

In early spring apply a general fertilizer such as Growmore at 60 g per sq. metre (2 oz per sq. yard) around the base of the plant. Water the surface and spread a 5 cm (2 in.) mulch of well-rotted compost around the stem.

Water thoroughly and regularly during drought conditions – this will not be necessary for established vines in most seasons. Feed dessert grapes with a high-potash liquid fertilizer from the time the fruits form until they begin to show colour.

Thinning may be necessary for dessert varieties but not for wine-making types. Bird protection will certainly be necessary when the grapes start to ripen – use netting to protect the bunches. Remove a few leaves as the bunches ripen in order to improve air circulation and to expose the fruit to sunlight.

## Picking and Storage

Colour is not a good guide to choosing the right time to pick – grapes may have to hang on the vine for some time to allow all the sugars to form – this may take a couple of weeks for an early variety or as much as a couple of months for the later varieties. Taste is the best guide.

Grapes are cut with a 'handle' – the whole bunch is removed by cutting the branch at about 5 cm (2 in.) on either side of the fruit stalk. Treat very gently. Grapes can be kept for 4–8 weeks by cutting the bunch with a length of branch attached and then putting one end in a jar of water. Store in a cool and dark place.

# MELONS

Melon growing once belonged inside heated greenhouses on the large estates of Victorian times, but it can now be practised much more widely. There are splendid modern varieties such as *Sweetheart* and *Ogen* which ripen well in many parts of the UK and there are now more garden greenhouses than ever before.

Still, you must be careful. Catalogues and books continue to talk about growing melons in the south without protection, but that can be risky. There are four acceptable sites – under cloches, inside a cold frame, in a polythene tunnel or in a greenhouse. Growing under cloches or inside a cold frame is usually successful in southern counties provided you choose an early variety and the summer is a sunny one. Growing in a cold greenhouse or tunnel is more reliable – choose an early or mid-season variety. The heated greenhouse is the true home for the melon in the UK – if you can maintain a minimum temperature of 18°C (65°F) then you can grow any of the varieties listed in the catalogues.

Melons are raised from seed at home, except for the mid-season netted variety *Galia* which is only sold as plug plants. If you succeed with greenhouse cucumbers then you should have no trouble with melons, although you will have to maintain a drier atmosphere. Early melons ripen in late July to mid-August; the mid-season ones are ready in late August; and the late varieties ripen at the beginning of September. Classification is notoriously difficult, but the arrangement shown below is accepted by most experts.

## Winter (Casaba) Melon

Large and oval, with hard, ribbed skin – yellow or green. The well-known *Honeydew* melon belongs here.

Flesh is pale green or white. It can be sweet and juicy. It is always crisp but is never fragrant.

A poor choice for a cold greenhouse or cold-frame cultivation.

## Netted (Musk) Melon

Small, round or oval, with smooth skin which bears fine or coarse fibrous netting – yellowish-green or buff.

Flesh is green, orange or pink. It is juicy and sweet and there is a distinct 'melony' aroma and taste.

Choose with care – some need a heated greenhouse. A poor choice for cold-frame cultivation.

## Cantaloupe Melon

Round, oval or flattened, with rough skin which is grooved but not netted – greyish-green or buff.

Flesh is green, orange or pink. It is juicy and sweet and there is a distinct 'peachy' aroma and taste.

The most popular group – a good choice for a cold greenhouse or cold-frame cultivation.

# Sowing and Planting

### Seed sowing

Sow seed in mid- or late April – sow a few extra pots in case of failure. Cover pots with glass and keep at a minimum of 18°C (65°F). Remove the glass after germination – maintain a minimum night temperature of 16°C (60°F). The seedlings are ready for planting out when there are 3–4 true leaves.

**2** Press in a single seed edgeways

**1** Fill 7–8 cm (3 in.) pot with seed- or multi-compost

### Cold-frame planting

Prepare the soil in April. Dig ½ bucketful of well-rotted manure or compost into the soil at the centre of the frame. Water if necessary. Close the lights. Harden off seedlings and plant out in late May. Close the lights after planting – cover surface for 7–14 days if sun is bright.

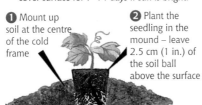

**1** Mount up soil at the centre of the cold frame

**2** Plant the seedling in the mound – leave 2.5 cm (1 in.) of the soil ball above the surface

**3** Do not firm. Water in – but keep water off the stem

### Greenhouse planting

You can plant into manure-enriched border soil or in compost-filled 25 cm (10 in.) pots, but growing-bag cultivation has become the most popular method. Put the bags in position at least 1 week before planting. Always plant at the side of the house and create a support system (see page 99) before planting.

**1** Place growing bag on the floor or staging

**2** Plant the seedling in the compost – 2 per bag. Leave 2.5 cm (1 in.) of the soil ball above the surface

**3** Do not firm. Water in – but keep water off the stem

Emir

No Name

Sweetheart

# Varieties A–Z

### Alvaro AGM

Possibly the best Charentais-type melon to ripen successfully outdoors in the UK. The fruits have a dark-striped, pale green skin which turns yellow as they ripen. The flesh is salmon-orange and is sweet, juicy and perfumed.

**Type:** Cantaloupe
**Fruit size:** Medium
**Skin colour:** Green-striped
**Flesh colour:** Orange
**Picking time:** Mid-season

### Blenheim Orange

An old favourite which keeps its place in some popular catalogues. This is one for the greenhouse, preferably a heated one. The flesh is occasionally more red than orange and the flavour is very good. Fruit is ready for picking in late August or early September.

**Type:** Netted
**Fruit size:** Large
**Skin colour:** Yellow
**Flesh colour:** Dark orange
**Picking time:** Late

### Emir AGM

This is an ideal choice for cold greenhouses, frames or under cloches. Fruits are round to oval shape, smooth, pale greenish-yellow, with sweet, juicy orange flesh. A melon that is tolerant of low temperatures and recommended for northern districts.

**Type:** Netted
**Fruit size:** Small
**Skin colour:** Pale green
**Flesh colour:** Orange
**Picking time:** Mid-season

### Galia

This well-known melon has a skin that is netted with gold, and firm, pale-green flesh with a spicy-sweet aroma. This is a very vigorous variety and needs plenty of room, but plants have good resistance to powdery mildew.

**Type:** Cantaloupe
**Fruit size:** Large
**Skin colour:** Yellow-green
**Flesh colour:** Green
**Picking time:** Mid-season

### Hero of Lockinge

A popular variety for growing in a heated house. Easy to recognize – a ball-like or broadly oval melon with golden, heavily netted skin and almost white flesh. The flavour is very good with a high reputation for sweetness. An old one but still widely recommended.

**Type:** Netted
**Fruit size:** Large
**Skin colour:** Orange
**Flesh colour:** Pale green
**Picking time:** Late

### No Name

Oddly named – and oddly described in some of the catalogues. The skin is creamy yellow with a variable amount of green. It is late fruiting and so you should pick another variety for growing in a frame or unheated house. The shape is oval and the flavour is excellent.

**Type:** Cantaloupe
**Fruit size:** Medium
**Skin colour:** Yellow or cream
**Flesh colour:** Orange
**Picking time:** Late

### Ogen AGM

Not quite as early as *Sweetheart*, nor is it as hardy. Still, it should be your choice for a cold frame or unheated house if you don't grow *Sweetheart*. The small, round, yellow fruits with green ribs are very popular – the firm flesh is juicy and sweet.

**Type:** Cantaloupe
**Fruit size:** Small
**Skin colour:** Yellow and green
**Flesh colour:** Pale green
**Picking time:** Early

### Sweetheart AGM

Everyone agreed – this *Charantais*-type melon is the one to grow in an unheated greenhouse or cold frame. It succeeds under cool conditions where all others would fail, and its flesh is both attractive and full of flavour. One drawback only – seeds are expensive.

**Type:** Cantaloupe
**Fruit size:** Medium
**Skin colour:** Grey-green
**Flesh colour:** Orange
**Picking time:** Early

# Training and Seasonal Care

## Greenhouse

Before planting it is necessary to create supports for the climbing plants. Stretch wires up to the eaves or the ridge; the stem will be about 1.8 m (6 ft) high. After planting tie a cane to the wires behind each plant.

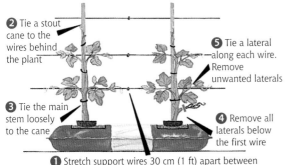

**2** Tie a stout cane to the wires behind the plant

**3** Tie the main stem loosely to the cane

**5** Tie a lateral along each wire. Remove unwanted laterals

**4** Remove all laterals below the first wire

**1** Stretch support wires 30 cm (1 ft) apart between vine eyes set 37 cm (15 in.) away from the glass

Remove the growing point when the greenhouse eaves or ridge has been reached. The stem should be about 1.8 m (6 ft high). Pinch out the growing tip of each lateral when 5 leaves have been produced – this induces the production of side-shoots. It is these side-shoots which produce the flowers.

The flowers are of 2 types. The males appear first (thin stalk behind the petals) and then the female flowers open (tiny melon behind the petals).

Hand pollination is necessary. Wait until 6 female flowers are open – each one on a separate lateral. Remove a mature male flower, fold back the petals and push gently into the heart of each female flower. Do this at midday. One male flower will fertilize 4 females. Thinning the fruit and stopping the side-shoots will be necessary when the melons are marble-sized – see below:

**1** Remove unwanted fruit to leave just one melon per lateral

**2** Pinch out stem 2 leaves beyond the fruit

You should now have 4–6 developing melons on each plant – remove any further shoots or flowers which appear.

Throughout the life of the plant careful watering is necessary. The compost should be thoroughly moist but never waterlogged – this can mean watering twice a day when the fruit is swelling, but you must keep water off the stems. Reduce watering when the melons start to ripen.

Regular feeding is also necessary. Begin feeding with a high-potash liquid feed when the fruits are the size of a golf ball and continue every 7–10 days until they start to ripen.

Gentle damping down and misting are desirable at most times, but should not take place during pollination time nor when the fruits start to colour. At this later stage the ventilators should be opened as a dry atmosphere is needed. Another requirement at the ripening stage is physical support – see below:

Net attached to wires to support fruit from tennis-ball size to maturity

## Cold frame

The appearance of new growth is an indication that the transplant is established – remove the shading material placed over the glass and start to increase ventilation gradually.

It is time to pinch out the growing point once the 5th true leaf has developed – this will induce side-shoots to form. About 3 weeks later select the 4 strongest shoots and train each one to a corner of the frame.

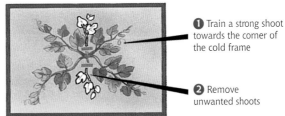

**1** Train a strong shoot towards the corner of the cold frame

**2** Remove unwanted shoots

The male and female flowers are borne separately. Once they appear the frame lights should be opened to allow insects to pollinate – replace at night. In wet and cold weather the lights should not be removed and it will be necessary to hand pollinate the flowers as for greenhouse-grown melons. Thinning the fruit and stopping the shoots will be necessary when the melons are marble-sized – see below:

**1** Remove unwanted fruit to leave just 1 melon per shoot

**2** Pinch out stem 2 leaves beyond the fruit

You should now have 4 melons on the plant – pinch out the tip of each of the 4 main shoots and remove any further shoots or flowers which appear.

Keep the soil moist but never saturated. Water regularly from the thinning stage illustrated above. Keep water off the stems as far as possible and reduce watering when the melons start to ripen.

Regular feeding is also necessary. Begin feeding with a high-potash liquid feed when the fruits are the size of a golf ball and continue every 7–10 days until they start to ripen.

Careful control of ventilation is necessary. Reduce ventilation in cold weather – increase ventilation in sunny weather and when fruits are ripening. Apply shading material to the lights in July and August. Another requirement at the ripening stage is physical support to keep the fruit off the soil which would cause it to rot – see below:

Tile or piece of wood placed under fruit from tennis-ball size to maturity

# Harvesting

Do not harvest melons until they are fully ripe. There are several indications – the end away from the stalk will give slightly if gently pressed and the end towards the stalk will start to develop a circular crack. Netted and cantaloupe types develop a characteristic aroma and the fruit when lifted parts readily from the stalk. Do not keep for more than a few days in the refrigerator.

# KIWI FRUIT

Once listed as Chinese gooseberry or Actinidia, the post-war success of the kiwi fruit has been phenomenal. Now you can buy them anywhere, but this is certainly not a plant you can grow anywhere. The shoot tips are susceptible to frost in the spring and prolonged sunshine is necessary to ensure proper ripening of the brown and furry fruit. If you live in the south and have a sheltered wall then it may be worth a try – a locality where sweet grapes grow successfully can usually be used with confidence for a kiwi fruit.

There is also a space problem – it is a rampant, twining climber which will reach 9 m (30 ft) or more if left unchecked. Of course this can be turned to your advantage if you have a large unsightly space or a pergola to clothe – the leaves are large and heart-shaped and the cream-coloured flowers are 3–4 cm (1½ in.) across.

Forget the recommendation to grow it under glass – this plant is far too vigorous for an average-sized greenhouse and it is not worth the trouble involved. Stick to outdoor growing and choose the female variety *Hayward* – undoubtedly the best and most reliable. Follow the rules for planting, pruning and care described for outdoor grapes on pages 93 and 96. There are a few differences – you will need to include a male plant such as *Tomuri* in each planting hole alongside the female one, and you will have to leave at least 3.6–4.5 m (12–15 ft) between these joint planting sites.

A new variety, *Jenny*, is self-fertile, so will produce a crop when planted on its own. It is less vigorous than many other varieties.

Hand pollination is recommended – follow the melon method on page 99. If the location is suitable you can expect about 9 kg (20 lb) of fruit from a mature plant, but you will have to wait for about 7 years before you get a worthwhile crop. The fruit is ready for picking when the brown surface gives slightly when gently pressed – this will usually be in October. Make sure all fruit is harvested before the first frosts arrive. Keep in a cool place for a month before use – kiwi fruit can be stored for a further month or two.

## Varieties A–Z

Hayward

Jenny

### Hayward

A female form producing large fruits 5–7 cm (2–3 in.) long x 5 cm (2 in.) across, with a pale brown skin covered with silky fine hairs. It is considered by many as the best variety for flavour. The vigour is a moderate 6–9 m (20–30 ft) and the late-flowering habit helps to avoid frost damage, making it a more reliable cropper.

**Type:** Female
**Fruit size:** Medium
**Skin colour:** Brown
**Flesh colour:** Green
**Picking time:** Late September–October

### Jenny

A new self-fertile variety producing sweet-flavoured, egg-shaped, greenish-brown fruit. The ideal choice for smaller fruit gardens, although it still needs room, reaching 6 m (20 ft) across and 3.6 m (12 ft) high.

**Type:** Male/female
**Fruit size:** Small
**Skin colour:** Green/brown
**Flesh colour:** Green
**Picking time:** September

### Tomuri

This male variety will not produce fruit, but is often grown as a pollinator for late-flowering female varieties such as *Hayward*. It is vigorous, often spreading to 9 m (30 ft) wide and 4.5 m (15 ft) high.

**Type:** Male
**Fruit size:** None
**Skin colour:** None
**Flesh colour:** None
**Picking time:** None

# CAPE GOOSEBERRIES

These are often grown as ornamental plants in gardens and you will find them labelled as Cape gooseberry, physalis or Chinese lantern in the supermarket. They are listed in many seed catalogues, but not always in the edible section. They are delicious served with the papery outer folded back and the colourful fruit part-dipped in chocolate.

It is grown as a half-hardy annual – outdoors in districts with a mild climate and sheltered situation, or under glass where the environment is less favourable. Be careful with this greenhouse recommendation – it can be grown in the same way as a tomato, but it is much too leafy for a small greenhouse.

Outdoor cultivation is quite straightforward if you have been successful in the past with outdoor tomatoes. You can expect about 0.5–1 kg (1–2 lb) of fruit from each plant and the lantern-like fruits provide an ornamental touch. Follow the general rules for tomato cultivation – sow seed in gentle heat in March and plant the seedlings under cloches in mid-May. Leave 75 cm (2½ ft) between the plants and remove cloches when the danger of frost is past. Insert 1.2 m (4 ft) stakes next to the young Cape gooseberries – tie the stems loosely to each support. Nip out the growing points when the plants are about 30 cm (1 ft) high and feed with a high-potash liquid fertilizer once the first fruits have formed. Watch for pests and spray accordingly.

The fruits in the greenhouse can be left on the plants to ripen – a golden shiny ball within a brown and papery husk. Outdoors you will have to pick at the unripe stage if frost threatens – leave the fruit to ripen on the windowsill.

Soft Fruit
Cape Gooseberries

## Variety
### Little Lanterns

**Type:** Half-hardy annual

**Fruit size:** Medium

**Skin colour:** Orange-yellow

**Flesh colour:** Pale-yellow

**Picking time:** Late September–October

This is a more compact form of Cape gooseberry to be found in some seed catalogues. The sweet fruits are pale green inside a green lantern-shaped husk and as the 'berries' ripen, the husk dries to a golden brown colour with a thin, papery texture. Grows to about 50 cm (20 in.) in height, with a less sprawling habit than other Cape gooseberries, making it a good choice for growing in containers

Little Lanterns

# The Major Pests and Diseases

| Strawberries | Raspberries | Blackberries and hybrid berries | Blackcurrants | Gooseberries | Red and white currants | Grapes | Melons |
|---|---|---|---|---|---|---|---|
| Greenfly | Greenfly | Big bud mite | Greenfly | Scale | Spider mite |
| Birds | Raspberry beetle | Greenfly | Greenfly | Gooseberry sawfly | Greenfly | Spider mite | Powdery mildew |
| Slugs | Raspberry moth | Raspberry beetle | Spider mite | Magpie moth | Capsid bug | Shanking | Soft rot |
| Virus | Birds | Birds | Birds | Magpie moth | Caterpillars | Powdery mildew | Mosaic virus |
| Strawberry mildew | Cane spot | Cane spot | Leaf spot | Birds | Leaf spot | Grey mould | |
| Grey mould | Spur blight | Spur blight | American mildew | Die-back | Coral spot | | |
| Red core | Grey mould | Grey mould | Grey mould | American mildew | | | |
| | Virus | Virus | Reversion virus | | | | |

## CAUSES OF POOR YIELDS

❧ **Impatience** Fruiting is light or absent and is often undesirable in the first season after planting. Strawberries and melons are an exception, but autumn-fruiting raspberries may not fruit heavily for 3 years after planting. With blackcurrants, gooseberries, red currants and white currants you will have to wait 4 to 5 years before a full crop is obtained.

❧ **Poor pruning and careless picking** The pruning instructions for grapes, blackberries, etc., may seem very severe as a large amount of growth is removed in winter. But it is necessary if you are to avoid an abundance of leafy growth and a dearth of fruit. Careless picking can also be a problem – do follow the instructions in the appropriate section. Tugging unripe fruit away from the stem can break fruiting spurs and damage stems – the remaining fruit may not ripen.

❧ **Birds** Perhaps the most serious soft-fruit problem – even a neglected plant may sometimes produce a good crop, but in nearly every case you can rely on birds to reduce or eliminate the ripening fruit. Netting is the only answer – see page 63.

Netting is required to keep birds away from buds and/or fruit. Polypropylene is preferred by many to nylon

❧ **Frost** A serious problem of soft fruit – see page 106.

❧ **Poor location or poor planting** Too much shade, poor soil, waterlogging, cold winds, too little time between digging and planting, dryness at the roots and planting at the wrong depth can all lead to disappointing crops.

❧ **Water and food shortage** Soft-fruit plants are often shallow rooting – regular and thorough watering is usually necessary during dry weather. Shortage of potash is a common cause of poor yields – feed as recommended. Apply a foliar feed containing seaweed extract if the leaves or stems have been damaged in any way.

❧ **Poor pollination** Rain, cold weather, strong winds or very dry air can all lead to defective pollination – see 'Blossom Drop' on page 106. Nature needs a helping hand in the case of indoor greenhouse-grown fruit.

Pollinate fruit grown under glass by hand. Tap flowers which are bisexual. With unisexual flowers (melons, kiwi fruits) transfer pollen to the stigma with a soft brush, cotton wool or by pushing a male flower into the female flower.

❧ **Pests and diseases** Pages 103–7 show that there are unfortunately a large number of problems which can beset the leaves, stems and fruits of soft fruit. Viruses are especially important and the effect on the infected plant gets steadily worse as time passes.

❧ **Soil diseases** There are a number of soil-borne diseases which can reduce the productiveness of affected plants – in severe attacks the cane or bush may be killed outright. Honey fungus (page 57) is well known, but there are also soft rot of melons (avoid overwatering), red core of strawberries (grow a resistant variety) and verticillium wilt of strawberries (grow a resistant variety).

❧ **Old age** Soft fruit has a limited life span in the garden, after which yields are disappointing. The average life expectancy of strawberries is 3 years, raspberries 10 years and gooseberries 10–15 years.

# HOLES IN LEAVES AND SHOOTS

There is a bewildering array of pests which can produce holes in the foliage of soft fruit. The most common culprits are illustrated on this page – others include the fruit tree tortrix moth (page 53), which spins leaves together, and the clay-coloured weevil (page 54), which occasionally attacks cane fruit. Many caterpillars can damage the leaves of soft fruit – not shown here are the winter moth, mottled umber moth and the vapourer moth (page 53). Slugs and snails (page 107) may attack strawberry leaves.

## Strawberry tortrix moth

A serious pest of strawberries in some areas of the UK. Several leaves are joined together by silken threads. Inside this protective cover the 6 mm (¼ in.) green caterpillars feed. These spun leaves should be picked off and destroyed. If spraying is necessary, apply thiacloprid, lambda-cyhalothrin or deltamethrin before flowering and repeat the spray after picking.

## Red-legged weevil

In parts of Britain this distinctive wingless weevil can be a serious pest of raspberries and strawberries. The weevils are rarely seen – they feed at night and drop to the ground when disturbed. Control is difficult – regular cultivation around the plants will help.

## Clearwing moth

Dead or dying shoots of currants and gooseberries may indicate the presence of the 1 cm (¾ in.) white caterpillars of this pest. The moths lay eggs in June on the shoots of the bushes and the caterpillars tunnel into the pith of the shoots. Cut back affected shoots to where the pith is no longer discoloured. Currant clearwing moth is not a widespread pest.

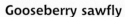

## Yellowtail moth

A hairy, colourful caterpillar which can cause a skin rash if handled. It is sometimes found on raspberry canes, where it can cause partial skeletonization of the foliage if present in sufficient numbers. The yellowtail moth, however, is a solitary feeder and so spraying with Thiacloprid or deltamethrin is rarely necessary. Just pick off and destroy the caterpillars.

## Capsid bug

Both apple capsid and the common green capsid can cause severe damage. The small green bugs puncture the leaf surface, producing reddish-brown spots. As the leaves expand, ragged, brown-edged holes are formed, often after the capsid bugs have left the plant. This pest can be a nuisance on all soft fruit, especially currants and gooseberries. Capsid bug is not an easy pest to control – spray with thiacloprid, lambda-cyhalothrin or deltamethrin when the first flowers are about to open and repeat the spray after the fruit has set.

## Raspberry moth

Dead or dying shoots of raspberries may indicate the presence of 8 mm (⅓ in.) red caterpillars of this pest. Young caterpillars hide in the soil or stakes during winter and move to the shoots in April, where they bore into and feed inside the pith. Cut out and burn withered shoots as soon as they are seen.

## Gooseberry sawfly

This serious pest of gooseberries and currants can cause complete defoliation in a severe attack. Identification is easy as the caterpillars generally feed round the edge of the leaf. Keep a careful watch for them from May onwards and spray with thiacloprid, lambda-cyhalothrin or deltamethrin at the first sign of damage. There can be up to 4 broods a year, so a second spray may be necessary later in the season.

*2.5 cm (1 in.) spotted caterpillar*

## Magpie moth

This distinctively coloured caterpillar feeds on gooseberry and currant bushes in spring and early summer and can cause severe defoliation. Fortunately it is now no longer a common pest and is only likely to be troublesome in small sheltered gardens where bushes are not sprayed. If only a few caterpillars appear, hand picking will give satisfactory control. Alternatively, spray with thiacloprid, lambda-cyhalothrin or deltamethrin when the first flowers are about to open.

*3 cm (1¼ in.) looper caterpillar*

# LEAF AND SHOOT PROBLEMS

Soft fruit can suffer from a wide range of leaf and shoot problems, as these two pages illustrate.

Not shown here are the clay-coloured weevil (page 54), crown gall (page 57) and cluster cup rust (page 107) on gooseberry leaves. Honey fungus (page 57) can affect gooseberries, raspberries, loganberries, strawberries, currants and vines.

Soft Fruit
Troubles

## Aphid

Many aphids attack soft fruit and their effect can be serious. In some cases there is severe leaf distortion, but the main danger is due to the viruses they carry. Currants are affected by the **currant blister aphid**, producing coloured blisters. Both **lettuce aphid** and **gooseberry aphid** cause severe leaf-curling on gooseberries. Two aphids attack raspberries – the **rubus aphid** and the **stem-coating raspberry aphid**. Strawberries can be attacked by other species – the **shallot aphid** which can cripple the plant, and the **strawberry aphid**. In all cases spray with thiacloprid, lambda-cyhalothrin or deltamethrin, pyrethrins or insecticidal soap as soon as the pests are noticed.

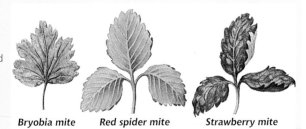

*Blister aphid*     *Leaf-curling aphid*     *Stem-coating aphid*

## Spider mite

Several mites attack soft fruit. General symptoms are pale or bronze-coloured leaves and minute 'spiders' on the undersurface. The **red spider mite** infests blackcurrants and strawberries in hot summers. On gooseberries the **bryobia mite** is a common pest, and defoliation is likely to occur in a severe attack. For both these mites spray with plant oils or extracts, fatty acids or urea/mineral lattice just after flowering if the weather is warm and settled. The **strawberry mite** is much more serious and much more difficult to control. Young leaves turn brown, older leaves are crinkled. Plants are stunted and the crown may die. Lift and burn infested plants.

*Bryobia mite*     *Red spider mite*     *Strawberry mite*

## Blackcurrant eelworm

The microscopic pests live on the young leaves and flowers within the bud. The infested buds shrivel and fail to open, so the branches remain bare in spring. There is no satisfactory method of controlling this pest, but luckily it is uncommon. Attacks are worst after a wet autumn. Cut off and burn bare branches.

## Strawberry eelworm

Several types of eelworm attack strawberries and identification is difficult. Leaf stalks may be abnormally long and turn red ('red plant disease') or leaf and flower stalks may be abnormally short and thickened. In all cases there is no cure – dig up and burn badly infested plants and do not replant for at least 5 years.

## Gall midge

A serious pest of blackcurrants, the small maggots feeding on the young leaves at the shoot tips. The young foliage becomes twisted, puckered and discoloured. Shoot growth may be checked. This is a difficult pest to control by spraying – no chemical sprays are approved for this purpose.

## Cane midge

The tiny pink maggots of the raspberry cane midge feed under the outer layer ('rind') of young canes. Direct injury is slight, but the damaged tissue is susceptible to attack by serious diseases which may kill the canes. Spray with thiacloprid, lambda-cyhalothrin or deltamethrin if growth splits appear on the stems – repeat the spray at the recommended intervals.

## Die-back

The dying back of odd branches of gooseberries is a frequent complaint – the cause is the grey mould fungus (page 107). A whole branch in full leaf suddenly dies; the foliage turns yellow and then withers and falls. Grey mould may be seen in the stem cracks. Remove affected branches promptly – paint cuts with a wound paint.

## Powdery mildew

Outdoor and greenhouse grapes and melons are attacked by this disease. White powdery patches appear on the leaves. On grapes this covering is often sparse so that browning may be more noticeable than the white mould. Burn diseased leaves. Apply an approved fungicide or a sulphur-based spray when disease is first seen and again at the recommended intervals.

## Cane spot

Raspberries and loganberries can be seriously affected. In early summer small purple spots appear on the canes. These enlarge to form shallow white pits with a purple border. The cane may be killed. Cut out badly diseased canes in autumn. Apply a copper-based spray when spots are seen on the canes and repeat at 14-day intervals.

## Brown scale

Several types of scale may be found on the stems of soft fruit. The tiny shell-like creatures do not move and are usually a sign of neglect. The commonest is the brown scale, which attacks currants, gooseberries and raspberries. Spray with thiacloprid, lambda-cyhalothrin or deltamethrin in March or September if the infestation is severe.

## American mildew

American mildew is a crippling disease of gooseberries. White powdery patches appear on the young leaves and shoots; later the fruit is affected (page 107). The mould changes from white to brown. This disease is encouraged by overcrowding – grow gooseberries on an open site and prune regularly. Cut off diseased branches in September. If the bushes have been badly attacked this summer apply myclobutanil next year when the first flowers open and repeat the spray twice at 14-day intervals. Blackcurrants are occasionally affected – spray with myclobutanil or potassium bicarbonate when the first spots of mould are seen.

## Reversion

This virus disease causes a change in leaf shape. The basal cleft disappears, the main lobe has less than 5 main veins and 10 serrations. The flower buds are red rather than grey. The virus is spread by the big bud mite (page 106) and reverted plants steadily degenerate. Destroy badly diseased plants – grow new bushes of certified stock on a fresh site.

*Healthy*   *Reverted*

cleft —   — no cleft

## Strawberry mildew

This is the most common disease of strawberries. It is easy to recognize – dark patches appear on the upper surface of the leaves, the infected foliage curling upwards to expose greyish mould patches on the undersurface. Fruit may also be affected (page 107). Spray with sulphur at 14-day intervals from the beginning of flowering until the fruit starts to colour. Cut off and burn leaves after harvesting.

## Leaf scorch

Leaves of all types of soft fruit may develop brown edges which are sometimes torn and curled. Growth is stunted and the fruit is small. The cause is potash deficiency – water the surrounding soil with a liquid potash-rich fertilizer and apply a dressing of general fertilizer in the spring.

## Rust

Yellow or black patches appear on the underside of the leaves in early summer – blackberries are occasionally badly infected. Collect and burn fallen leaves in autumn – spraying is rarely worthwhile and crop yields are hardly affected.

## Leaf spot

Leaf spot is the most serious disease of blackcurrants. Gooseberries may also be attacked. Brown spots appear on the leaves and early defoliation takes place. The bushes are weakened and next year's crop will be reduced. Leaf spot is worst in a wet season – spraying is a worthwhile precaution. Apply a copper-based spray at the first sign of leaf spotting and repeat at 14-day intervals if weather remains damp. Pick off diseased leaves – rake up and burn all fallen leaves. Strawberry leaves sometimes develop red spots; spraying is rarely necessary.

## Virus

Virus diseases are a major problem of raspberries and strawberries. Aphids and other insects are the carriers, and once the plants are infected there is nothing you can do to save them. **Mosaic** is a serious disorder of raspberries. Strawberries are affected by several viruses – **crinkle** occurs in late spring, **yellow edge** in autumn and **arabis mosaic** in either spring or autumn. The only control is to remove and burn diseased plants.

*Mosaic*

*Yellow edge*

*Crinkle*

## Spur blight

Purplish patches appear around the buds on raspberry and loganberry canes in early autumn. These patches turn silvery and the buds are killed. Reduce overcrowding by removing unwanted canes. Cut out and burn diseased canes as soon as the purple patches start to appear. To prevent attacks spray with a copper-based spray at 14-day intervals from bud burst until the blossom appears.

## Frost: cane fruit

Some or all of the buds on fruit canes may fail to open in the spring. If the area around the withered bud is discoloured, the cause is raspberry spur blight. If the surrounding bark appears normal then the most likely reason for bud failure is frost. This damage usually occurs when there has been an early mild spell which causes the sap to rise, followed by a frost as the buds were about to burst.

## Blossom drop

If the plants are healthy and not suffering from drought then there are two possible causes of lack of fruit set. Either frost occurred when the bushes were in flower or pollination was defective due to wet and cold weather or very dry air at blossom time.

## Big bud mite

This gall mite is by far the most serious pest of blackcurrants. Red currants and gooseberries are occasionally attacked. In late spring or early summer these microscopic mites are carried by wind or other insects to the bushes. By July they have entered the buds, which swell and become less pointed. These 'big buds' eventually wither. Inspect the bushes carefully in winter. Pick off enlarged buds; dig out and burn badly infested plants

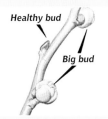

*Healthy bud*

*Big bud*

## Strawberry blossom weevil

Both strawberries and raspberries can be attacked by this pest. The flower stalk is partially severed after eggs have been laid inside the flower bud. The unopened flower then withers and may fall. Blossom weevil damage is sometimes confused with a strawberry rhynchites attack; blossom weevil is easily recognized by its greyish-black body. The damage is rarely serious – if spraying is necessary use thiacloprid or deltamethrin late in the evening.

## Strawberry rhynchites

Both strawberries and raspberries can be attacked by this pest. In May and June the stalks of blossom trusses are punctured, the unopened buds wither and occasionally fall. Rhynchites damage is sometimes confused with a blossom weevil attack; strawberry rhynchites is easily recognized by its greenish metallic body. The damage is rarely serious – if spraying is necessary use thiacloprid or deltamethrin late in the evening.

## Frost: strawberries

Open strawberry blooms are extremely sensitive to frost. The effect is unmistakable – the petals remain apparently unharmed but the central core turns black. Nothing can be done at this stage – the flowers wither and fall. The time to take action is before the frost occurs – cover the plants with cloches or newspaper if frost is forecast at blossom time.

## Birds

Birds are a menace to all types of soft fruit when the berries or currants are ripening. At the bud stage gooseberries are most at risk; in many gardens some or all of the dormant buds are stripped off by bullfinches during the winter months. The only really satisfactory answer is netting. Make sure the base of the net is well secured. If netting is not practical, wind cotton thread through the branches and delay pruning until the buds are breaking in the spring.

## Coral spot

A problem for tree fruit (see page 57) but also a very serious problem for soft fruits, especially currants and gooseberries. Coral spot appears on the surface of twigs and shoots before invading live tissue and killing whole branches on established bushes. Cut away and destroy all diseased areas.

# FRUIT PROBLEMS

Soft fruit is sweet and succulent, so birds are a major menace. Slugs, snails and black beetles devour fruit which is close to the ground and in a wet summer grey mould can destroy a crop.

Dropping of small fruitlets occurs occasionally; in blackcurrants it is known as 'running off' and is due to frost damage or poor pollination. In raspberries this trouble is often due to damaged fruiting spurs.

## Sun scald and shanking

Sunken patches on grapes. In shanking the fruit stalks shrivel and there is overall shrinkage of the fruit – the cause is over-cropping or under-watering. Scald occurs in hot weather – stalks do not shrivel and shrinkage is at the top of the fruit. Apply shading material to the glass and ventilate in sunny weather. Remove damaged fruit.

## American mildew

This form of powdery mildew begins as a white coating on gooseberries and currants, but later it becomes a brown felt over the surface. Bushes should be pruned regularly and infected shoots removed in September. Apply myclobutanil when the first flowers open. Repeat twice at 14-day intervals.

## Slugs, snails and beetles

Large holes in strawberries may be due to slugs and snails (look for slime trails) or strawberry ground beetles. Attacks are worst in enclosed gardens. Use metaldehyde or ferric (iron) phosphate pellets scattered thinly around the plants. Alternatively, use a biological control containing parasitic eelworms. Remove garden rubbish.

## Birds and squirrels

Animals are usually the cause when fruit is entirely devoured. Birds, such as blackbirds, will eat strawberries, and squirrels are rapidly becoming the major pest of this crop in many areas. Deterrents are of limited use and the only answer is netting. Ensure the base of the net is well secured.

## Raspberry beetle

This is the most serious pest of raspberries, loganberries and blackberries as the 6 mm (¼ in.) white grubs can soon ruin the crop. Spray raspberries when the first fruits start to turn pink – loganberries when the petals have fallen. Use a contact insecticide, based on pyrethrins, insecticidal soap, fatty acids or urea/mineral lattice – check label for any period needed before picking.

## Vine powdery mildew

This disease is common on both outdoor and greenhouse grapes. The fruit is covered with a white powdery mould, and cracking may take place. Prune the vines thoroughly in autumn, following the instructions on page 96. Burn all infected leaves and wood. Dust with sulphur when the first flowers open. Repeat twice at 14-day intervals.

## Cluster cup rust

Minute yellow-edged pits ('cluster cups') appear on the large orange patches covering infected fruit. This easily recognizable disease occurs only where sedges are growing nearby. Badly affected bushes should be replaced – spread them widely apart. Chemical spraying is rarely effective and so is not worthwhile.

## Strawberry seed beetle

These beetles are about 1 cm (½ in.) long, greyish, and very active. They bite at the seeds and the attached flesh so that the fruit is disfigured. Keep down weeds, clear away dead leaves and remove garden rubbish. Rather similar damage can be caused by birds. Insecticides are no longer available.

## Strawberry mildew

This powdery mildew is a serious foliage disease of strawberries, which can attack the berries. Diseased fruit are dull-coloured and sometimes shrivelled. In a severe outbreak they are unfit for use. Spray with urea/mineral lattice when disease is first seen. At replanting time choose a modern variety which has good mildew resistance.

## Grey mould (botrytis)

This fluffy mould is destructive to raspberries, strawberries, grapes and currants in a wet summer. Remove diseased fruits and mouldy plant material immediately. The answer in a humid season used to be to spray with a fungicide before the disease appears, but there are no longer any approved chemicals for the control of this disease.

## Chapter 4

# CONTAINER-GROWN FRUIT

Growing fruit in pots or containers has a number of benefits and offers several advantages over growing plants in the open ground, especially if you are gardening in a limited space. Using containers gives you a measure of control over the size and growth of your plants by a combination of pruning and container size, but it is important to choose appropriate plants for your situation.

Fruit trees, bushes and canes can all grow in containers in very small spaces, which is ideal for homes with small gardens and flats with just a balcony. Strawberries will grow in pots or troughs on a windowsill or flight of steps. It is usual to keep pots and containers close to the house so that watering is easier, and this allows you to observe how the plants are growing. It is also convenient for harvesting the fruit when it is ready.

Growing in containers allows you to offer extra protection to the plants in winter by moving them to a more sheltered position where they can avoid the worst of the frosty and windy conditions. It is also an asset if you wish to grow plants that may struggle in your garden soil. For example, blueberries need acidic conditions but can be grown in pots even if your garden soil is unsuitable for them.

Containers are mobile and can be moved to the fore when they are looking attractive or when you are waiting to harvest the fruit, then moved back out of the way when they are resting. They can also be taken with you if you move house.

## Choosing Compost

While it might be tempting to think that because garden soil is suitable for the plants in the garden it is also suitable for container-grown trees, you always get better results with bought compost mixed to a known formula. These composts contain controlled-release fertilizers, which will feed the plants over a set period of time before they need to be replenished. They are also designed to provide good drainage as well as holding moisture so that the plant does not dry out too easily. John Innes loam-based composts and peat-based multi-purpose composts are the best choices.

### TREE FRUIT

Most tree fruits tend to perform better when grown in loam-based compost, such as John Innes No. 3. The weight of the loam provides added stability, which is important for taller plants.

### SOFT FRUIT

Peat-based compost with added shredded bark is most commonly used for blueberries, due to their specific low pH requirements. Multi-purpose peat-based compost is preferred for strawberries, the plants of which have a much shorter lifespan than other fruits, at around 3 years.

# Choosing a Container

Pots and containers come in all shapes and sizes. They are usually described either by the volume (capacity) of compost they can hold in litres, or by diameter measured across the upper rim of the container.

The container may be round, square or rectangular, with vertical or tapered sides. For taller plants, which might be unstable in windy conditions, a container with vertical sides where the width is the same at the top and bottom (a cylinder or cube) will offer more stability than a container with a base that is narrower than the top.

The shape of a container can also influence the growth and management of the plant inside it. Wide, shallow containers offer a large surface area of compost and very little depth (in relation to the surface area), so the moisture loss due to surface evaporation and drainage is much greater than that from a deep, narrow container, even if both hold the same volume of compost.

## TREE FRUIT

For new fruit trees, the most suitably sized pots or patio containers will be those with a top diameter of at least 60 cm (2 ft). For a square container this equates to having sides of about 45 cm (18 in.) and a volume of around 60 litres (13 gallons).

## SOFT FRUIT

The types of container used for soft fruit can vary considerably, mainly because of the different growth habits of the plants involved. Strawberries are shallow-rooted but have a spreading habit, while raspberries have a shallow root system and a tall, upright habit. Currants and blueberries need deep containers to accommodate their roots and bushy growth.

Sturdy containers with good drainage are essential for soft fruit (they hate waterlogging) and light containers are best for plants such as strawberries and raspberries because they can be moved under protection to start cropping earlier in the season or continue later at the end.

*Tall containers are ideal for strawberries. The fruit is kept clean, visible and easy to pick.*

## Types of Container

**Plastic** containers offer the greatest range of shapes and sizes, as well as being light and durable. Look for those made with thicker-grade plastic as they offer more protection to plant roots in winter. Black ones are best avoided for a sunny spot, as they absorb heat which can reduce root growth.

**Terracotta** has a natural, traditional appearance, but these pots are heavier than plastic and may break if dropped. They breathe, meaning air can pass in through the sides, but also that moisture can be lost; to reduce moisture loss, line the inside walls of the container with plastic sheeting before planting. Look for a frost-resistant guarantee, as many do not fare well in frosty weather and can crack.

**Wooden** containers (especially half barrels) are ideal for larger-growing and taller plants because of the stability they offer. They also have a natural appearance. Like terracotta, moisture can be lost through the sides, so line the inside walls of the container with plastic sheeting before planting to minimize this. Their size and weight also mean that they are difficult to move once they have been planted.

**Metal** containers look attractive, but if they are placed in a sunny spot the metal will conduct heat very efficiently and the heat can cause damage to the roots, reducing plant growth. In winter, the roots of the plants can also suffer, as the thin metal offers very little protection from the cold.

*Troughs work well for some fruits such as blueberries, but the shallow depth and large surface area of compost makes them very prone to drying out quickly.*

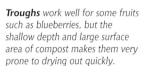

*Cylinder containers with a broad base are very stable, making them perfect for taller plants.*

*'Herb pots' are ideal for strawberries. The small surface area of exposed compost makes them lose moisture more slowly.*

*'Vase' containers are better for deep-rooted plants such as figs, or for those where the fruit tends to hang from the plant.*

**109**

# Tree Fruit

When growing fruit trees in pots, you will need to feed regularly with a balanced fertilizer during spring and summer to replace nutrients as they are used up from the compost. After flowering and during fruit swell, administer a high-potash feed every 2 weeks. Make sure the compost stays moist in hot weather, as drying out may be detrimental to fruit production.

## CHOOSING THE RIGHT ROOTSTOCK AND VARIETY

The vigour of the rootstock will determine the habit and ultimate height of a fruit tree, but while a dwarfing rootstock would seem to be the obvious choice, this is not always the case. Dwarfing rootstocks, by their very nature, can lack vigour and are not always suitable for growing within the restrictions of a container. Trees grown on semi-vigorous rootstocks usually prove to be a better choice.

Choosing fruit tree varieties that are slower-growing or not vigorous in combination with a more vigorous rootstock is often the best idea, with the aim of restricting the size of the tree to about 1.8 m (6 ft), usually as a dwarf bush or dwarf pyramid tree.

When selecting the plants you wish to grow, always look on the label – you should see the cultivar name with the rootstock printed next to it, such as Apple 'Sunset' M9.

### Rootstocks

**Apple** M9, M26 (M27 is too dwarfing)
**Pear** Quince C
**Apricot** St Julien A or Torinel
**Plum, damson, peach, nectarine** Pixy or St Julien A
**Cherry** Colt or Gisela 5
**Figs** are grown on their own roots

*The graft union, where the variety and rootstock are joined, often shows as a large swelling because of the unequal growth rates of the different parts of the tree.*

*Trees growing in containers can be attractive as well as productive.*

### Tree fruits for containers

| Apples | Pears | Plums | Cherries |
|---|---|---|---|
| Blenheim Orange | Beth | Blue Tit | Lapins |
| Discovery | Beurre Hardy | Opal | Stella |
| Egremont Russet | Concorde | Victoria | Sunburst |
| Falstaff | Conference | **Peaches** | |
| Fiesta | Doyenne du | Duke of York | **Figs** |
| Greensleeves | Comice | Peregrine | Brown Turkey |
| James Grieve | Glou Morceau | Rochester | Brunswick |
| Katy | Goldember | | White |
| Lane's Prince Albert | **Apricots** | **Nectarines** | Marseilles |
| Pixie | Goldcot | Early Rivers | |
| Spartan | Moorpark | Fantasia | |
| Sunset | Tomcot | Nectarella | |

## PATIO FRUIT TREES

Traditionally, fruit trees sold for growing in containers were reliable varieties that had been grown in gardens and orchards. They were considered suitable for containers because they were less vigorous. Their size and shape could be controlled or restricted by using rootstocks and container size to limit their growth. In recent years, however, plant breeders have set out to produce genuinely dwarf trees which would naturally stay small enough to grow well in containers.

There is an increasing range of such dwarf fruit trees to choose from – they are ideal for smaller gardens. Their restricted size does not affect fruit size or yield. An average tree in this range is about 1.2 m (4 ft) wide and 1.5 m (5 ft) high after 7 years, if well cared for.

The modern method of growing fruit trees in containers is to train them as columns or vertical cordons. They are also known as Minarettes, Pillarettes or Supercolumns.

### Patio trees for containers

| Apple | Pear | Peach | Apricot |
|---|---|---|---|
| Garden Sun Red | Garden Gem | Amber | Apricompakt |
| Lilly | Garden Pearl | Crimson Bonfire | Aprigold |
| Sally | Lilliput | Diamond | **Figs** |
| **Plum** | **Cherry** | **Nectarine** | Panachee |
| Black Amber | Garden Bing | Rubis | |
| | Maynard | | |

# Soft Fruit

## FEEDING

Most soft fruit plants are quite hungry feeders and will need additional nutrients to those already in the potting compost. This can be used to the gardener's advantage, as adjusting the fertilizer can influence what the plant will do. Adding higher levels of phosphates and potash (in the form of tomato feed) can be used to encourage root development, flowering and fruiting. Applying a balanced fertilizer such as Liquid Growmore after fruiting can be used to promote extra shoot growth to produce better, stronger canes and stems for the following year.

## MOVING

The great advantage of growing plants in containers is increased mobility. Fruiting plants can be moved into a more sheltered position to provide added protection during the winter months. Even better, moving the plants provides an opportunity to extend the fruiting season; for example, relocating strawberries into a protected area can make them fruit several weeks earlier than they would do outdoors in the garden and, if late-cropping varieties are at risk of frost damage in the autumn, they can be moved under protection to continue fruiting.

It must be remembered that soft fruits are hardy plants and will need to spend part of the winter period outdoors. They need a period of cold temperatures for their flowers to develop naturally in order to crop well the following year.

# Care and Maintenance

## WATERING

Watering is the biggest issue when choosing to grow fruit in containers. These plants have a high demand for water, particularly when the fruit is developing, but they rely completely on the gardener for its supply. Although most of these plants will be dormant during the winter months, they will need watering once or twice a week during the rest of the year and in hot, dry weather it may be once or even twice a day.

Plants suffering from a shortage of water often succumb to powdery mildew and may even start shedding the developing fruits if they are too dry.

The smaller and shallower a container, the more likely it is to need frequent watering, and wide, shallow containers have the greatest rate of moisture loss.

## WINTER PROTECTION

Most hardy fruits can survive the winter outdoors, because the leaves, stems and branches have evolved to cope with lower temperatures. However, plants growing in the garden have their roots insulated by the soil around them, whereas a plant in a container is more exposed. Frost can penetrate the sides and base, killing young roots close to the sides of the container. Wrapping several layers of protective bubble polythene around the container will usually give the roots sufficient protection, especially if the plants have been moved to a more sheltered position.

---

### Soft fruits for containers

Strawberries have long been popular fruits for containers, but a wide range of other bush and cane fruits can be successfully grown this way too.

#### Strawberries

Strawberries are one of the easiest fruits to grow in pots and there are a range of patio containers to suit any space. The advantage of container strawberries is that they are grown above ground, making it easier to keep slugs off and the fruits clean, as well as make picking easier.

**Suitable varieties:** *Elsanta, Malling Centenary, Pandora, Pegasus, Sweetheart*

#### Raspberries

Raspberries can be grown in containers on the patio as long as the container is of a suitable size, about 60 cm (2 ft) in diameter. Long-cane raspberries are ideal for this, as these are full-length canes, which will produce flowers and fruit from the side-shoots in the first season after planting. Planting 6 canes around the edge of the container and tying the tops together to form a 'wigwam' keeps the plants tidy and easy to manage.

**Suitable varieties:** *Erika, Joan J, Malling Admiral, Malling Jewel, Octavia, Tulameen*

#### Blueberries

Most varieties of blueberry can be grown in a container, but it must be at least 30 cm (12 in.) in diameter and the compost must be ericaceous as they need acidic conditions. Water with rainwater rather than tap water, and feed only with an ericaceous fertilizer.

**Suitable varieties:** *Bluecrop, Chandler, Herbert*

---

## RE-POTTING

It is not essential to move the plants to a larger container each year. You can remove the top 7–10 cm (3–4 in.) of compost and replace it with fresh compost to top up the nutrients for several years before re-potting becomes necessary. This has the added bonus of removing any weeds and weed seeds that have got into the top surface layer at the same time.

After tree, cane and bush fruits have reached their final size of container, this 'top dressing' should be carried out each spring just as the new growth starts to appear and it is worth replenishing a larger portion of the compost every 2–3 years.

Established fruit trees may be removed from their containers and their roots pruned during winter before re-potting in fresh compost. This usually involves pruning the roots back by about a quarter to encourage the plant to continue growing whilst preventing it getting too big.

## HORIZONTAL STAKING

When growing tree fruit in containers, it can be difficult to keep taller plants stable after re-potting, as the compost may not be firm enough to hold a support stake without rocking in windy conditions. This can be dealt with by using horizontal staking to secure the stem of the plant (plus a vertical stake if necessary) within the container.

# Chapter 5

# FRUIT FOR ALL SEASONS

The average gardener with a vegetable plot and a large freezer can satisfy most of the family's needs for greens and root vegetables throughout the year. The situation is different with fruit. Home-grown produce does have a role to play, but it can provide only a minor part of a family's annual consumption.

A minor part, but an important one. On page 3 the special role of growing fruit was highlighted – the ability to eat fruit fresh from the bush or cane, the chance to grow top-flavour varieties that commercial growers ignore and the opportunity to grow some unusual types, such as hybrid berries, which are in the catalogues but rarely seen in the shops.

However, for several reasons we must still turn to the shop and supermarket for most of our year-round fruit needs. Modern research has shown that fruit is vital for good health, but some of the basic fruits in our diet will not grow in Britain – grapefruits, oranges, lemons and bananas are notable examples. Several other types *can* be grown here with care and protection, but in practice very little is actually grown at home, although for some types that is increasing – for example blueberries, peaches, nectarines, melons and grapes. But even with our most popular home-grown favourites which do flourish in our climate there is a problem – apples, pears and strawberries cannot be successfully stored at home for year-round use in the same way as broad beans and Brussels sprouts – so if we want to enjoy fruit out of season, we have to shop for it. Also, we must face the fact that flavour is often confused with freshness by gardeners and the claim that *all* fruits taste better when grown at home and eaten straight from the plant is simply not true. While it may be so for fruit that is completely at home in Britain, such as the raspberry and many apple varieties, it is not true for fruit that needs warm sun on its back. Examples are melons, kiwi fruits, grapes and some imported apples, such as *Braeburn*, *Golden Delicious*, *Granny Smith* and *Pink Lady*.

The supermarket has become the main source of supply – the range is large and the quality is generally high. But pre-packed fruit is nearly always more expensive – sometimes much more expensive – than fruit sold loose, so obviously there is still a place for the greengrocer, the farm shop and the market stall.

Over time our tastes have changed, as more people holiday abroad than ever before and sample cuisine from all parts of the globe. Not only has this influenced the meals we cook and eat, but also the different fruits and vegetables we try. People look for these 'exotic' fruits once they return home and now expect to find at least some of them for sale. Listed on the following pages are some of the more exotic types of fruit that can be bought in the UK, some well known, but others distinctly unusual and unable to be grown in this country.

## Shopping Rules

Shopping wisely involves learning a few rules:

### Look at the fruit
• Buy fresh, sound produce – no mouldy spots, no leaking punnets, no squashed soft fruit.

### Look at the label
• This will be stuck on each individual pre-pack or on each individual fruit, or it will appear as a display card next to fruit that is sold loose. For all popular fruit it is now a legal requirement for this label or display card to tell you the type, the country of origin and the class of the produce.
• Class I means that the physical quality is good and there are no important defects.
• Class II indicates that the quality is reasonable but the fruit is deficient in one or two respects – the shape or colour may be less than good or there may be small blemishes or marks present.

• If quality deteriorates the label should be changed or the fruit should be taken off the shelves. In addition, the variety must be shown on the label or display card for many Class I fruits.
• Remember that the class is not an indication of eating quality or flavour, but tells customers the standard to which the fruit has been graded – a Class I melon can be unripe and tasteless. This can be the case with other fruits as well, particularly if they have been transported long distances. Ripe fruit does not travel well, so is harvested before it is fully ripe and ripened during transport. Telling when a fruit is in peak condition is not always easy – look up the notes in this section for the individual fruit.

### Buy for a few days only and eat as soon as possible
• Storage life decreases as temperature increases and the ripening process accelerates, so do keep fruit cool.

**Wash bought fruit well** and peel where necessary.

## ASIAN PEAR

You will not find this one in the cookery books but it is beginning to appear in the larger supermarkets from November to April. It doesn't look like a pear – it has the appearance of an apple with a pale yellow skin. The flesh is crisp and distinctly aromatic, and the refreshing taste is quite unique – nothing like an apple or pear. Eat it fresh – don't waste its unusual flavour in a fruit salad. Lightly poach in red wine if you must cook them.

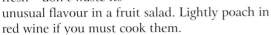

## AVOCADO

Avocado or alligator pear has a dark green or black-skinned, fleshy body that may be pear-shaped, egg-shaped or spherical. They are usually picked when fully developed but while unripe to ripen after harvesting. The flesh is greenish-cream in colour, usually becoming paler as the fruit becomes fully ripe. Remove the stone by cutting the fruit lengthwise and scoop out the flesh with a spoon. Once ripe, keep them in the fridge; the flesh discolours rapidly on exposure to air but this can be prevented by treating with lemon juice.

## BABACO

Search for this one between October and March – it is worth looking for. It is large, looking like a fluted marrow and weighing up to 1 kg (2 lb). Babaco was discovered in Ecuador in the 1920s and has been developed commercially by New Zealand farmers. The fruit is green when unripe and bright yellow when ripe and ready for eating. The centre is hollow with practically no seeds – the skin is edible. Slice and eat like star fruit (see page 120).

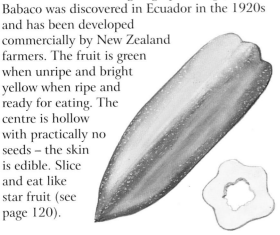

## BANANA

The ordinary dessert banana is far too well known to require detailed description. Available all year round, the fruit is curved, 15–30 cm (6–12 in.) long and is ready for eating when the yellow skin is flecked with brown. Always store bananas at room temperature – they rapidly blacken in a refrigerator.

The other large group of bananas are the plantains, also known as cooking or Jamaican bananas. These are larger and duller-skinned than the ordinary ones and have an insipid flavour as the sugar content is low. The plantain is much used in West Indian and African cooking and you will find this fruit in shops catering for these ethnic groups. Other types of banana are hard to find. There are lady's fingers (bright yellow, small, very sweet and thin-skinned) and the red banana (dull red skin and pink flesh).

Bananas provide a snack *par excellence* and also are a basic ingredient for fruit salad. Place in lemon or other acidic juice after peeling and slicing to prevent browning. There are numerous other ways of using bananas. They can be boiled, baked, flambéed or fried in sweet batter as fritters.

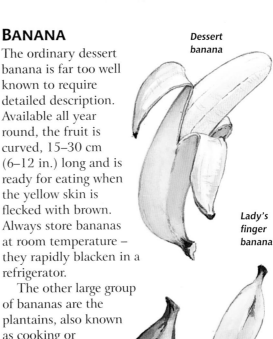

*Dessert banana*

*Lady's finger banana*

*Red banana*

*Plantain*

## BER

Here is a truly unusual one which has not been taken up by the supermarkets. You will have to go to an Asian shop during winter if you want to try this fruit. It is eaten fresh; the white flesh around the stone is crisp and refreshing. It has long been known in the tropics – alternative names are Indian plum and Chinese apple. It has been used for hundreds of years as a tonic and cure for chest complaints – another alternative name is jujube.

## CITRON

It looks like an unusually large lemon, but unlike all other citrus fruits the citron is used for its thick peel rather than its flesh or juice. This peel is candied for use in cake and biscuit making and decoration. The strips of peel are simmered in boiling water for a few minutes and then washed in cold water. Repeat this process several times. The next stage is to simmer the peel in a sugar syrup for about 3 hours. Dry in a warm place and store in an air-tight jar.

## CLOUDBERRY, CRANBERRY AND DEWBERRY

There is no precise definition for this group, which includes several berry-type fruits which are gradually increasing popularity and use. The cloudberry is popular in Scandinavia, as the ripe fruits are tart but rich in vitamin C. They are often made into jams, juices, tarts and liqueurs. Several single-berry types are becoming very popular as shoppers look for more variety. The red cranberry is very tart – cook slowly with sugar until the berries burst. This tangy fruit is frequently used as an accompaniment to turkey. The dewberry is similar to the blackberry, but it is grey, smaller and more delicately flavoured – use fresh or cook with apples for tarts, jam, etc.

**Cloudberry  Cranberry  Dewberry**

## COCONUT

Actually a fruit, rather than a true nut. The hard, brown 'stone' of a ripe coconut is covered in a thin layer of fibre. The fruit is broadly oval in shape with three pores or 'eyes' clearly visible at one end. It is hollow inside and filled with a sweet liquid known as coconut milk. The inside walls of the shell contain a layer of edible, white 'flesh' up to 2.5 cm (1 in.) thick, which is rich in fibre and nutrients and can be eaten fresh, dried or in cooked dishes. A full-sized coconut weighs about 1.44 kg (3.2 lb).

## CUSTARD APPLE

These are fruits of the anona tree – related to the pineapple and breadfruit. You may find one or more types in larger supermarkets in late autumn or winter, but you would have to go to the West Indies to find them in all shops. The white or creamy flesh is soft like custard and the shape is sometimes apple-like – hence the common name. All contain numerous black seeds.

*Sugar Apple* (*Sweetsop*) is a yellowish-green, round or oval fruit made up of large and fleshy scales which burst open when ripe. The sweet flesh has a banana-like flavour. *Cherimoya* is larger, round or pear-shaped with a smoother pale green skin. The aromatic creamy flesh has a pineapple flavour. *Soursop* is the giant of the group – up to 30 cm (1 ft long) and weighing 4.5 kg (10 lb). The thin skin has rows of spines and the flavour is more acidic and refreshing than the others. *Bullock's Heart* is heart-shaped with a rather insipid taste. *Atemoya* is a US-bred hybrid.

Buy when the skin is slightly soft to the touch. Keep in the refrigerator and cut in half as a dessert. Eat the pulp with a spoon – the seeds are inedible – or remove seeds and serve the mashed pulp in a bowl.

**Sugar Apple**

**Cherimoya**

**Soursop**

## DATE

Dates can be bought fresh or dried and are available all year round. The most popular form is dried and packed as a compressed block with a stalk in a long box. The fruit should be plump, smooth-skinned and free from sugar crystals. To remove skin, cut off one end and squeeze the other – the skinned date will pop out. Remove the stone by cutting the date lengthwise – serve alone or fill with cream cheese or marzipan. Dates are also used in baking.

## DRAGON FRUIT

Dragon fruit or strawberry pear is an oval-shaped fruit up to 10 cm (4 in.) long, bright red in colour and coated with green-tipped scales. Inside, the white, juicy flesh contains large numbers of tiny edible black seeds embedded in a fragrant, juicy, sweet pink or white flesh. The fruit can be eaten chilled, in fruit salads (add lime to bring out the flavour) or used to flavour drinks and pastries. The flower buds can be cooked and eaten as vegetables.

## DURIAN

You might find this strange fruit in canned form but not in the fresh state, and that is a good thing. The soft flesh within the spiky rind is delicious. But the smell of the fruit is halfway between raw sewage and bad eggs. Surprisingly, these large oval fruits of 2.25–4.5 kg (5–10 lb) are popular in the markets of Malaysia and Indonesia and are obviously an acquired taste. If offered one, keep in a polythene bag before serving.

## FEIJOA

The feijoa or pineapple guava has skin the texture and colour of an avocado, but the soft flesh and the seeds in a gelatinous pulp indicate that it is closely related to the guava. Popular in some parts of the US and increasing in popularity in Britain, with fruits imported from New Zealand. Buy when the skin is slightly soft to the touch and eat immediately. Cut in half and scoop out with a spoon. Do not eat the skin.

## GOJI BERRY

The goji berry or wolfberry produces a bright orange-red fruit 1–2 cm (½–¾ in.) in diameter and 3 cm (1½ in.) long, with large numbers of tiny yellow seeds inside. The red berries, which can be eaten fresh, cooked or dried, contain high levels of protein and vitamin C, which has led to them being hailed as a 'super fruit'. Although many claims have been made regarding the health benefits of goji berries, medical experts stress that there is little clinical evidence to support this.

## GRAPEFRUIT

The largest and one of the most popular citrus fruits – an essential start to the day for many people. The usual 'white' types have pale yellow, juicy flesh with a few large pips and a thick skin. Pink grapefruit is much sweeter – better for eating fresh but less suitable for squeezing. The pinks have rose-coloured flesh but the surface may be yellow, pink or pale green. Look for shiny and heavy fruit which gives slightly when pressed. Grapefruit will keep for a few weeks in a cool place.

## GUAVA

You will find these fruits on the shelves mainly in autumn and winter – as small as an egg or apple-sized, oval or round, yellow or green, rough- or smooth-skinned. The refreshingly tart flesh surrounds a soft pulp containing numerous seeds. The lemon guava has near-white flesh and the strawberry guava has pink flesh. The strong musky smell is not particularly pleasant – store away from other foods. Cut in half and eat with a spoon – with a large fruit remove the pulp and eat the flesh.

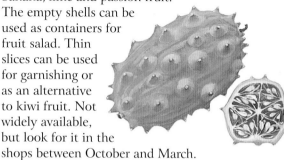

## HONEYBERRY

This edible honeysuckle has blue fruits which are about 1 cm (½ in.) in diameter and approximately 3 cm (1½ in.) long, although shape and size can vary considerably. The fruits are covered in a white bloom, while the flesh turns burgundy-red and becomes sweet and tasty. Changing the name to honeyberry made this fruit much more sought-after in the US. The fruits are very rich in vitamin C; they can be eaten fresh or used to make jams and jellies.

## JACKFRUIT

The jackfruit (or jakfruit) is the largest fruit which is grown commercially, sometimes reaching 36 kg (80 lb) or more. However, the ones seen in the UK weigh only about 1.5 kg (3–4 lb) – they are oval with a greenish skin which is hard, thick and warty. The aroma is strong when the fruit is ripe – the succulent flesh is eaten fresh and the flavour may be delicate or insipid – you can't tell from the outside. Large seeds are embedded in the flesh – these can be roasted and served like chestnuts.

## KIWANO

Another interesting addition from the New Zealand fruit growers – a brand name for the horned melon. The outer case is bright orange and spiky; within is sweet green pulp with lots of seeds. Eat the pulp with a spoon straight from the shell – the taste is claimed to be a blend of banana, lime and passion fruit. The empty shells can be used as containers for fruit salad. Thin slices can be used for garnishing or as an alternative to kiwi fruit. Not widely available, but look for it in the shops between October and March.

## KUMQUAT

A tiny, orange-like fruit which is usually oval in shape – a round version is sometimes available. All you have to do is to wash the fruit and eat it whole – cut it in half if you want to. The skin is sweet and the flesh is refreshingly tart. Another use is to garnish fruit salads and cakes with thin slices – best of all is to preserve the fruits in brandy. Store fresh fruit in the refrigerator for up to a week. Despite their appearance, kumquats are not related to citrus.

## LEMON

One of the most versatile fruits of all – the basic souring agent for the British cook. Lemonade, pies, garnishing, marmalade, salad dressing, sauces, desserts – the uses are truly extensive. Choose fruit which is hard and glossy. Lemons which are rather small and heavy are usually juicier than large, knobbly ones, which are mainly grown for their peel. Lemons will keep for several weeks in a fridge – wrapped in polythene they will keep for a couple of months in a cool place. Cut lemons, however, should be used as soon as possible.

# LIME

Once preferred to lemons in Britain, the lime is back in favour. It contains more citric acid than the lemon, so reduce the amount you would normally add when using as a lemon substitute. The shape of the two fruits is similar but lime has a thinner skin, which changes from green to yellow with age. The flesh, however, always remains pale green. Caribbean limes are larger and brighter-coloured than Indian ones – both types will stay fresh for many weeks in the refrigerator.

# LIMETTA

Many of the fruits in this section are everyday items – others are not yet widely known but have appeared in recent years on supermarket shelves. The limetta (sweet lemon) is one of the few real rarities which have been included. The fruit is rounder than the ordinary lemon and the taste is quite different – the juice is tangy, not strongly acidic, so that it can be squeezed like grapefruit to provide a refreshing drink.

# LONGAN

These oriental fruits occasionally appear in the shops when the lychee season is over. The two are very closely related but there are some distinct differences. The skin of the longan is much less knobbly and the flesh is more aromatic, more translucent and the flavour is less acidic and more grape-like. Do not buy if the skin is shrivelled. Prepare and use longans in just the same way as lychees (see opposite).

# LOQUAT

The loquat or Japanese medlar has the size and shape of an elongated small plum and the skin colour and texture of an apricot. The juicy flesh, which is tart rather than sweet, surrounds 2 or 3 large stones. To serve, wash, and remove the stones. The skin can be peeled away if desired. You will find them in the shops in spring and summer – pick ones with firm and unblemished skins. Recommended for poaching – remove skin after cooking.

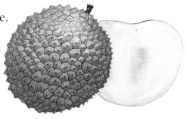

# LYCHEE

Try fresh lychees if you only know this fruit from the canned version served in Chinese restaurants. The tough, knobbly skin is easily peeled away to reveal the juicy white flesh, within which is a large brown stone. Serve cold – eat the flesh, discard the stone. When buying, pick out lychees with skin which has a distinctly pink flush to the background brown colour, and there should be no sign of shrivelling. Store in a polythene bag in the refrigerator.

# MANGO

A commonplace sight nowadays in most supermarkets – for many years it was a rarity. Size, shape and skin colour are extremely variable – there are green, yellow, orange, red and purple varieties. With the popular types, green usually indicates that the fruit is not yet ripe. The flesh is always bright orange with a central large stone. Prepare by slicing the flesh away on either side of the stone. Serve on its own, with ice cream or in fruit salad.

## MANGOSTEEN

A round fruit, the size of an apple or orange, with a thick, leathery, purplish-brown skin and the calyx firmly attached at the top. Break open the outer coating by hand. Inside is a ring of fleshy segments – white, juicy and fragrant, with a flavour rather similar to lychees. Each segment contains a couple of seeds.

## ORANGE

Several types of **sweet orange** are available. *Jaffa* is large, thick-skinned, almost seedless and very juicy. It is popular, as is the smooth-skinned *Navel* with its characteristic dimple at one end. *Valencia* is thin-skinned, perfectly round and small. It is sweet, virtually seedless and very juicy. The small *Blood Orange* has reddish, sweet flesh and red-flecked skin.

*Jaffa*

   **Tangerines** (or mandarins – the terms are interchangeable) are types of orange with loose peel that pulls away easily. There are several varieties and hybrids.

   The basic tangerine is small and sweet with bright orange skin and many pips. The smallest variety is the *Wilkins* with yellowish-orange skin and lots of pips. The *Satsuma* is larger and more popular – originally from Japan but now grown in Spain. The sweet flesh is almost seedless; the skin is loose. Similar but bigger is the *Clementine*, with firm, deep orange skin.

*Navel*

   There are several tangerine x sweet orange hybrids. *Dancy* is American with a rather tart flavour. *Topaz* has a richer, sweeter taste. *Temple* is easily recognized by its red skin and is usually sold as an orange. The largest is the *Ortanique*, which looks like a flattened orange.

*Blood*

*Clementine*

   *Tangelos* are tangerine x grapefruit hybrids. *Minneolas* have deep orange skin, few seeds and a sharp flavour.

   Tangerines are at their best at Christmas. Eat fresh, use as garnish or in fruit salad. They also make excellent marmalade.

*Ortanique*

## PASSION FRUIT

The passion fruit or purple granadilla is about the same size and shape as a plum, but the purple skin is leathery and wrinkled. The flesh is yellowish-orange with a distinct aroma and flavour. Cut the fruit in half and scoop out the tangy flesh and the crunchy seeds. A popular alternative use is to strain the flesh and incorporate the juice in fruit punches or cocktails.

The sweet granadilla (larger, orange and smooth) is sometimes available.

## PAW PAW

Shaped like a large pear, paw paw (or papaya) can weigh up to 1 kg (2 lb). The smooth skin turns from green to yellow or pale orange when ripe – the surface is usually slightly speckled with green. The thick layer of melon-like flesh under the skin is pinkish-orange and the central cavity is filled with black seeds. Prepare as if it was a melon. Cut in half, remove the seeds and slice into wedges. Sprinkle with lemon or lime juice. It has been known for some people to be allergic to paw paw juice.

## PERSIMMON

Persimmon (or kaki fruit) looks like a large orange-red tomato with a prominent greenish calyx at the top. The flesh of unripe fruit is unpleasant and bitter when eaten – make sure persimmon is ripe before serving. Place the fruit inside a bag and keep in a warm place until the tough skin is translucent and the surface gives when gently pressed. Cut off the top and scoop out the sweet, jelly-like flesh with a spoon. Discard the large seeds.

## PINEAPPLE

Once a food for the very wealthy, this juicy, sweet-acidic fruit is commonly available in most supermarkets and greengrocers. Pineapple must be ripe when you buy it – it will not ripen in your home. Smell it – there should be a strong pineapple aroma. The leaves should be stiff and fresh-looking – on a ripe fruit a central leaf when gently tugged will pull away quite easily. The easiest way to serve is to cut it into slices, then into pieces, or serve as rings. Remove the outer rind and then the inner core.

## PITANGA

The pitanga or Surinam cherry is truly a rarity, although one of its close relatives, the clove, was a mainstay of the spice trade for hundreds of years. The fruits are red and are borne in clusters like cherries, but the similarity stops there. The surface of each fruit is deeply ridged and the flesh contains several pips. Pitanga is grown in New Zealand and Israel – the two enterprising pioneers of exotic fruit. It can be eaten fresh, but is more often used in preserves.

## POMEGRANATE

Some of the exotic fruits described in this chapter were unknown in Britain before the war. The pomegranate is different – it has declined in popularity with most fruit going to make juice. The orange-sized fruit has a leathery skin which should be red-tinged and shiny. Neither the skin nor the inner pith is edible – the juice is packed in sacs around each seed. Cut off the top – eat with a spoon, sucking the juicy flesh from each sac and spitting out the seed.

## POMELO

The pomelo or shaddock looks like a large, misshapen grapefruit with a yellow or yellowish-green skin. A more recent arrival to our shops – surprisingly, it is the parent of the grapefruit and not a new variety. Remove the thick peel and strip away the bitter skin around each segment. These sections are served as an alternative to grapefruit – the texture is coarser but the flavour is less bitter. The peel is recommended for candying.

## PRICKLY PEAR

Prickly pear, cactus fruit, Indian fig – all alternative names for the fruit of the Opuntia cactus. Oval, about 7–8 cm (3 in.) long, the skin is cream, orange or pink. Take care – the surface has small groups of tiny barbed spines. Never handle this fruit. Cut in half and peel with a knife and fork before scooping out the flesh. This sweet pulp bears masses of hard but edible seeds. Ripe fruit can be kept for 2 weeks in the refrigerator.

## RAMBUTAN

A close relative of the lychee, but you wouldn't think so from its appearance on the shelf. The plum-sized fruit is covered with a mass of curved spines, giving it a fearsome appearance. These soft spines are green at first, but turn rich ruby red when the fruit is ripe. Break the shiny case open and the translucent, succulent flesh within immediately reveals its close link to the better-known lychee. Use in the same way – remove the flesh from the central shiny stone.

## SAPODILLA

The sapodilla or naseberry has the appearance of a stalked new potato. The skin is brown and rather rough – an unappetizing cover for a delectable fruit. Within, the flesh is soft, granular and fragrant. This pale brown pulp has a taste like banana-flavoured brown sugar – obviously a fruit well worth trying. Store until it is ripe – the skin should be yielding and should show no green when scratched. Chill, peel and cut in half to remove the dangerous hooked pips.

## SHARON FRUIT

The persimmon (page 118) never became popular in Britain. It has been overtaken by a variety bred in Israel, and named the Sharon fruit. Rather more orange than red, this variety is seedless and can be eaten whilst it is still quite firm. The skin is edible, so it's just a matter of wash, slice and eat. The flavour is a blend of melon and peach – rather sweeter than the parent persimmon. Mostly available between November and January.

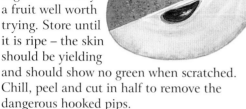

## STAR FRUIT

In recent years the star fruit, star apple or carambola has appeared on the shelves of larger supermarkets and as a garnish on desserts and fruit salads served in restaurants. The waxy fruit, approximately 10 cm (4 in.) long, is distinctly fluted. When the yellow translucent skin and crunchy flesh is sliced, 5-pointed stars are obtained. Remove the pips and serve fresh. Sometimes sweet, sometimes distinctly tart – it is difficult to predict.

## SWEETIE

Another novelty bred in Israel, mainly available in supermarkets between October and March. It looks like an unripe grapefruit, but the skin remains completely or partly green even when ripe. It is seedless and free from acidity. The promoters recommend a novel way of serving sweetie. Cut off the top and bottom and then cut the fruit into 8 segments. Each segment is held by the thick peel and the sweet flesh eaten. Of course, you can eat it in the traditional grapefruit way.

## TAMARILLO

The tamarillo (other names: tree tomato, Java plum) is always included in the fruit section, but it is close to a tomato in appearance, taste and use. It is large – about 12–15 cm (5 in. long) and oval; the orange-red skin is tough and inedible. When ripe the flesh gives when gently pressed. Cut in half and scoop out the flesh with a spoon – the flavour is sweet and acidic. Excellent when baked, stewed or grilled as a vegetable. Peel before use.

## UGLI

An ugly fruit, hence the name. This tangerine x grapefruit hybrid exhibits characteristics from both parents. It is the size of a grapefruit and the pinkish-yellow flesh is more tart than a tangerine. The greenish-orange wrinkled peel, however, is distinctly tangerine-like – it is loose and easily pulled away to reveal the segments. The skin around these segments should be removed before serving. Buy fruit which feels quite firm and is free from bruises. Store in a cool place.

# GLOSSARY

**Acid soil** A soil which contains no free lime and has a pH of less than 6.5.

**Adult foliage** Leaves on a mature branch which differ in shape and size from the *juvenile foliage*.

**Adventitious bud** A growth bud which appears abnormally on the stem, compared to a normal *axillary bud*.

**Alkaline soil** A soil which has a pH of more than 7.3. Other terms are chalky soil and limy soil.

**Alternate** Leaves or buds which arise first on one side of the stem and then on the other. Compare *opposite*.

**Annual** A plant that grows from seed, flowers and dies all within one season.

**Anther** The part of the flower which produces pollen. It is the upper section of the *stamen*.

**Apex** The tip of a stem or other plant organ.

**Apical bud** A growth bud at the tip of a stem.

**Armed** Bearing strong thorns.

**Awl-shaped** A narrow leaf which tapers to a stiff point.

**Axil** The angle between the upper surface of the leaf stalk and the stem that carries it.

**Axillary bud** A growth bud which appears in the angle between a leaf and the stem.

**Bare-rooted** A plant dug up at the nursery and sold with no soil around its roots.

**Bark-ringing** The removal of a strip of bark around a tree in order to reduce vigour and increase fruitfulness. Compare *root pruning*.

**Basal cluster** A group of leaves forming a complete or partly complete ring at the base of a shoot.

**Bedding plant** A plant which is bedded out in quantity and has only a limited life outdoors.

**Berry** A pulpy fruit bearing several or many seeds.

**Biennial** A plant which grows from seed, producing stems and leaves in the first season and flowering in the next.

**Biennial bearer** A tree which regularly bears a good crop of blossom or fruit every other year, with little or no crop in the intervening year.

**Bisexual** A flower bearing both male and female reproductive organs – compare *dioecious* and *monoecious*.

**Bleeding** The abundant loss of sap from severed plant tissues.

**Bloom** Two meanings – either a fine powdery coating or a flower.

**Bole** An alternative name for the *trunk* of a tree.

**Branched head** The collection of branches at the head of a tree in which there is no *central leader*.

**Breaking** The development of shoots as a result of removing wood above an *axillary bud*.

**Breaking bud** A bud which has started to open.

**Breastwood** Shoots which grow forward from a *supported tree*.

**Bud** A flower bud is the unopened bloom. A growth bud or *eye* is a condensed shoot.

**Budding** A form of *grafting* where a single bud rather than a shoot is used as the *scion*.

**Bush** A tree pruned to give 60–75 cm (2–2½ ft) of clear stem.

**Calcareous** Chalky or limy soil.

**Calcifuge** A plant which will not thrive in *alkaline soil*.

**Callus** The scar tissue which forms over a pruning cut or at the base of a cutting.

**Calyx** The whorl of *sepals* which protect the unopened flower bud.

**Cambium** A thin layer of living cells between the bark and the wood.

**Cane** A long and slender shoot arising from the *crown* of the plant.

**Canker** A diseased and discoloured area on the stem.

**Central leader** The main vertical leader at the centre of the tree – the *trunk*.

**Chelate** An organic chemical which can supply nutrients to a plant in a soil that would normally lock up the plant-feeding element or elements in question.

**Chlorosis** An abnormal yellowing or blanching of the leaves due to lack of chlorophyll.

**Clingstone** A type of *stone fruit* in which the flesh is tightly attached to the stone. Compare *freestone*.

**Clone** A group of identical plants produced by *vegetative reproduction* from a single plant.

**Compatible** The *scion* (or bud) and the *rootstock* are able to unite successfully after grafting.

**Compost** Two meanings – either decomposed vegetable or animal matter for incorporation in the soil or a potting/cutting mixture made from peat ('soilless compost') or sterilized soil ('loam compost') plus other materials such as sand, lime and fertilizer.

**Compound** A type of leaf which is composed of several *leaflets*.

**Container grown** A plant for sale that has been grown in a pot (or similar container) throughout its life. Compare *field grown*.

**Containerized** A plant that has been grown in a nursery field and transferred into a pot (or similar container) only for sale.

**Cordon** A supported type of tree trained vertically or obliquely and with a main stem or stems pruned to produce *spurs*.

**Corolla** The whorl of petals within the *calyx* of the flower.

**Cross** The offspring arising from cross-pollination.

**Crotch** The angle between large branches or between a branch and the *trunk*.

**Crown** The bottom part of a *herbaceous* plant from which roots grow downwards and the shoots arise.

**Culinary fruit** Fruit which is cooked or processed by jam-making, etc., rather than being eaten fresh. Compare *dessert fruit*.

**Cultivar** Short for 'cultivated variety', it is a variety which originated in cultivation and not in the wild. Strictly speaking, all modern varieties are cultivars, but the more familiar term '*variety*' is used in this book.

**Current year's growth** Shoot development which has been made this season. Also known as *this year's growth* or current season's growth – the wood produced is called *first-year* or *new wood*.

**Cutting** A piece of stem cut from a plant and used for propagation.

**Deblossoming** The removal of flowers or flower trusses to avoid overcrowding in mature plants or to direct the vigour into vegetative growth in young plants.

**Deciduous** A plant which loses its leaves at the end of the growing season.

**Dehorning** The cutting back of stout branches to a point at which another branch arises.

**Deshooting** The removal of unwanted shoots while they are still small and soft.

**Dessert fruit** Fruit which is suitable for eating in the fresh uncooked state. Compare *culinary fruit*.

**Dioecious** A plant which bears either male or female flowers – e.g. many kiwi fruit varieties. Compare *monoecious*.

**Disbudding** The removal of surplus buds before they have started to develop.

**Dormant period** The time when the plant has naturally stopped growing due to low temperature and short day-length.

**Double working** The grafting of a variety on to a compatible *rootstock*, which in turn is grafted on to the desired rootstock. This system is used when the variety and the rootstock are *incompatible*.

**Downy** Covered with soft hairs.

**Dwarf bush** A tree pruned to give 45–60 cm (1½ –2 ft) of clear stem.

**Dwarf pyramid** A tree pruned to form a broadly conical shape with a *central leader* about 2.1 m (7 ft) high.

**Espalier** A supported type of tree trained to form a vertical main stem with pairs of branches stretched horizontally to form a series of tiers.

**Extension growth** A shoot which has arisen from an *apical bud*.

**Eye** A dormant growth bud.

**Fan** A supported type of tree trained to form a series of main branches spreading like the spokes of a wheel or the ribs of a fan.

**Feathered maiden** A 1-year-old tree bearing *lateral* shoots. Compare *maiden whip*.

**Fertilization** The application of *pollen* to the *stigma* to induce the production of seed. An essential step in hybridization.

**Field grown** A plant for sale that has been grown in a nursery bed or open area (field). Compare *container grown*.

**Filament** The supporting column of the *anther*. It is the lower part of the *stamen*.

**First-year wood** See *current year's growth*.

**Floricane** Plants which produce stems that grow for one year before producing *lateral* (side) shoots bearing flowers and fruit the following year. Compare *primocane*.

**Flush** A crop of flowers or fruit, which may or may not be followed by other flushes.

**Foliar feed** A fertilizer capable of being sprayed on and absorbed by the leaves.

**Framework** The basic woody skeleton or the main branches of the tree or bush.

**Free-standing tree** A tree grown in the open without any horizontal supports. Compare *supported tree*.

**Freestone** A type of *stone fruit* in which the flesh is free from or loosely attached to the stone. Compare *clingstone*.

**Frost pocket** An area where cold air is trapped during winter and in which tender plants are in much greater danger.

**Fruit** The seed together with the structure that bears or contains it.

**Fruit bud** Large rounded bud producing blossom then fruit.

**Fungicide** A chemical used to control diseases caused by fungi.

**Fungus** A primitive form of plant life which is the most common cause of infectious disease – e.g. mildew and rust.

**Genus** A group of closely related plants containing one or more species.

**Germination** The emergence of the root and shoot from the seed.

**Grafting** The process of joining a shoot or bud of one plant (the *scion*) on to the stem and root system (the *rootstock*) of another.

**Growth bud** Small flat bud which gives rise to a shoot.

**Half-hardy** A plant which will not survive if the temperature falls below 0°C.

**Half-standard** A tree pruned to give 1.2–1.35 m (4–4½ ft) of clear stem.

**Hand pollination** Using a soft brush or similar to transfer *pollen* from the male to the female parts of a flower.

**Hardening-off** The process of gradually acclimatizing a plant raised under warm conditions to the environment it will have to withstand outdoors.

**Hardy** A plant which will survive outdoors over winter without protection.

**Heading back** Pruning the *central leader* of a *maiden*. Removing the top growth of a *rootstock* after a successful budding *take*.

**Heeling-in** The temporary planting of a new tree or shrub pending suitable weather conditions for permanent planting.

**Herbaceous** A plant which does not form permanent woody stems.

**Humus** Term popularly (but not correctly) applied to partly decomposed organic matter in the soil. Actually humus is the jelly-like end-product which coats the soil particles.

**Hybrid** Plants with parents which are genetically distinct. The parent plants may be different *cultivars*, *varieties* or *species*.

**Incompatible** The *scion* (or bud) and the *rootstock* are not able to unite successfully after grafting. *Double working* is required.

**Inorganic** A chemical or fertilizer not obtained from a source which is or has been alive.

**Insecticide** A chemical used to control insect pests.

**Internode** The part of the stem between one *node* and another.

**Interstock** A section of stem inserted between the *scion* (or bud) and the *rootstock* if they are not able to unite successfully during grafting. See *double working* and *incompatible*.

**Joint** See *node*.

**Juvenile foliage** Young leaves which differ in shape and size from the *adult foliage*.

**Lateral** A side-shoot borne by a *leader*.

**Leader** A main branch. The leading shoot determines the general direction of growth of the tree.

**Leaflet** One of the parts of a *compound* leaf.

**Maiden** A 1-year-old tree or shrub.

**Maiden whip** A 1-year-old tree with no *lateral* shoots. Compare *feathered maiden*.

**Monoecious** A plant which bears both male and female flowers. Compare *dioecious*.

**Mulch** A layer of bulky organic material placed around the stems.

**Mutation** A sudden change in the genetic make-up of a plant, leading to a new feature. This new feature can be inherited.

**Native** A *species* which grows wild in this country and was not introduced by man.

**Neutral** Neither acid nor alkaline – pH 6.5–7.3.

**New wood** See *current year's growth*.

**Node** The point on the stem at which a leaf or bud is attached.

**Nut** A one-seeded hard fruit which does not split when ripe.

**Offset** Young plant which arises naturally near to or at the base of the parent plant and is easily separated.

**Old wood** Stem growth produced before the current season.

**Opposite** Leaves or buds which are borne in pairs along the stem. Compare *alternate*.

**Organic** A chemical or fertilizer which is obtained from a source which is or has been alive.

**Ovule** The part of the female organ of the flower which turns into a seed after fertilization.

**Patio trees** Dwarf fruit trees developed for growing in containers and restricted spaces.

**Peat** Plant matter in an arrested state of decay obtained from bogs or heathland.

**Perennial** A plant capable of living for at least 3 years.

**Petal** One of the divisions of the *corolla* – generally the showy part of the flower.

**pH** A measure of acidity and alkalinity. Below pH 6.5 is acid, above pH 7.3 is alkaline.

**Pinching out** The removal between finger and thumb of the growing tip of the stem.

**Pollen** The yellow dust produced by the *anthers*. It is the male element which fertilizes the *ovule*.

**Pollination** The application of *pollen* to the *stigma* of the flower.

**Pome fruit** A hardy tree which bears fleshy fruit with several small seeds in the central cavity.

**Pricking out** The first planting out of a seedling or rooted cutting into another container or outdoor bed.

**Primary branches** The first branches to develop on the main stem or *trunk*.

**Primocane** Plants which produce stems that yield fruit in their first year of growth. See *floricane*.

**Propagation** The multiplication of plants.

**Pruning** The removal of parts of the plant in order to improve its shape and/or performance.

**Pyramid** A tree pruned to form a broadly conical shape with a *central leader* over 2.1 m (7 ft) high.

**Regulatory pruning** The removal of weak, diseased and overcrowded branches.

**Renewal pruning** The removal of wood to ensure a steady supply of new shoots.

**Reversion** Two meanings – either a *sport* which goes back to the colour or growth habit of its parent, or a cultivated variety which is outgrown by suckers arising from the *rootstock*.

**Rod** The main stem of a grapevine.

**Root pruning** The removal of part of the root system in order to reduce vigour and increase fruitfulness. Compare *bark-ringing*.

**Rootstock** See *grafting*.

**Runner** A shoot which grows along the soil surface, rooting at intervals.

**Russet** A brown roughening or scurfing on the surface of some fruits, especially apples and pears.

**Scion** See *grafting*.

**Secondary branches** The branches which develop on a *primary branch*.

**Self-fertile** A flower which can be successfully pollinated by its own *pollen* or from pollen produced by other flowers on the tree or bush.

**Sepal** One of the divisions of the *calyx*.

**Snag** A short stump left after careless *pruning*.

**Species** Plants which are genetically similar and which reproduce exactly when self-fertilized.

**Spindlebush** A form of *pyramid* tree in which the side branches are permanently tied down.

**Spit** The depth of the spade blade – usually about 23 cm (9 in.).

**Sport** A plant which shows a marked and inheritable change from its parent – a *mutation*.

**Spur** A short and slow-growing branch which bears fruit buds.

**Spur-bearer** A tree which bears all or most of its fruit on *spurs*.

**Stamen** The male organ of a flower, consisting of the *anther* and *filament*.

**Standard** A tree pruned to give 1.8–2.1 m (6–7 ft) of clear stem.

**Stigma** The part of the female organ of the flower which catches the *pollen*.

**Stone fruit** A hardy tree which bears fleshy fruit with a large hard seed at the centre.

**Stopping** See *pinching out*.

**Strig** Fruit cluster of red currant, white currant or blackcurrant.

**Strike** The successful outcome of taking cuttings – cuttings 'strike' whereas grafts *take*.

**Sub-lateral** A side-shoot on a *lateral*.

**Sucker** A shoot growing from the *rootstock*.

**Supported tree** A tree which is grown against a fence, framework of wire or a wall. Compare *free-standing tree*.

**Synonym** An alternative plant name.

**Systemic** A pesticide which goes inside the plant and travels in the sap stream.

**Take** The successful outcome of budding – grafts 'take' whereas cuttings 'strike'.

**Terminal bud** A bud at the end of a 1-year-old shoot.

**Thinning** Reducing the number of fruits in a cluster to increase the size and yield of the remaining fruits. Also reducing the density of branches to improve air flow and reduce the chance of fungal disease.

**This year's growth** See *current year's growth*.

**Tilth** The crumbly structure of surface soil.

**Tip-bearer** A tree which bears most of its fruit at the tips of 1-year-old shoots.

**Tipping** Removal of the growing point by means of secateurs rather than *pinching out*.

**Top dressing** An application of fertilizer to feed established plants.

**Trained tree** A tree which has been trained on to a framework of supports. See *cordon, espalier* and *fan*.

**Transplanting** Movement of a plant from one site to another.

**Trunk** The woody *central leader* of a tree.

**Truss** A cluster of fruit or flowers.

**Union** The junction after grafting of the *scion* and *rootstock*. Also the name of the raised scar where this junction has taken place.

**Unisexual** A flower of one sex only – see *monoecious* and *dioecious*.

**Variety** Strictly speaking, a naturally occurring variation of a *species* (see *cultivar*).

**Vegetative growth** The development of stems and leaves, as opposed to blossom and fruit development.

**Vegetative reproduction** Budding, cuttings, division, grafting and layering, as distinct from sexual reproduction by seeds.

**Virus** An organism which is too small to be seen through a microscope and which is capable of causing serious malformation of a plant.

**Water shoot** A vigorous soft shoot which is unbranched and unfruitful.

# INDEX OF FRUIT

Index

Index